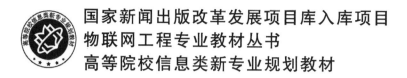

国家新闻出版改革发展项目库入库项目
物联网工程专业教材丛书
高等院校信息类新专业规划教材

物联网专业英语

主　编　许　可
副主编　刘　佳　任春蕾　王姗姗　赵　欣

北京邮电大学出版社
www.buptpress.com

内 容 简 介

本教材从物联网的基本技术、应用、政策三个方面精选15篇文章作为课文,并结合产出导向的教育理念,在技能讲解和习题部分设计了学术、生活及工作三个场景,以及综合阅读、词汇、听说、写作四大板块,以全方位提高读者的语言技能。本教材在场景设计之下融合物联网的专业知识,使读者能够掌握精确的行业术语,深入思考物联网技术为生活和社会治理带来的深刻变革,并使用英语灵活讨论行业相关内容。本教材可作为物联网、电子信息、微电子、计算机等相关专业的教材或培训用书。

图书在版编目(CIP)数据

物联网专业英语 / 许可主编. -- 北京:北京邮电大学出版社,2020.9(2023.3重印)
ISBN 978-7-5635-6191-9

Ⅰ. ①物… Ⅱ. ①许… Ⅲ. ①物联网—英语—教材 Ⅳ. ①TP393.4②TP18

中国版本图书馆 CIP 数据核字(2020)第 148667 号

策划编辑:姚　顺　刘纳新　　责任编辑:马晓仟　　封面设计:七星博纳

出版发行:北京邮电大学出版社
社　　址:北京市海淀区西土城路 10 号
邮政编码:100876
发 行 部:电话:010-62282185　传真:010-62283578
E-mail:publish@bupt.edu.cn
经　　销:各地新华书店
印　　刷:保定市中画美凯印刷有限公司
开　　本:787 mm×1 092 mm　1/16
印　　张:13.5
字　　数:348 千字
版　　次:2020 年 9 月第 1 版
印　　次:2023 年 3 月第 2 次印刷

ISBN 978-7-5635-6191-9　　　　　　　　　　　　　　　　　定价:38.00 元

· 如有印装质量问题,请与北京邮电大学出版社发行部联系 ·

物联网工程专业教材丛书

顾问委员会

邓中亮　李书芳　黄　辉　程晋格　曾庆生　任立刚　方　娟

编委会

总 主 编：张锦南
副总主编：袁学光

编　　委：颜　鑫　左　勇　卢向群　许　可　张　博
　　　　　张锦南　袁学光　张阳安　黄伟佳　陈保儒

总 策 划：姚　顺
秘 书 长：刘纳新

前言

长久以来,我国的英语教育一直为人们所诟病,其中一个原因就是其仅把重点放在语言技能的训练上,很少涉及专业知识,即在英语课上所讨论和练习的话题通常都是日常生活中的话题。这对于初高中学生学习英语来说可能还够用,但对于大学生学习英语来说,仅对日常话题的讨论是远远不够的。学生在学习过程中和在未来的工作中更需要的是将英语与本专业的知识相结合,能够使用英语在专业内进行沟通,或者进行学术研究。因此在我国英语教学的诸多侧重点中,专门用途英语变得热门起来。

编者的设计理念是使本教材成为拥有物联网专业背景的综合英语教程,因此本教材涉及听、说、读、写及词汇训练的各个方面,并基于产出导向法,编者在技能讲解和习题部分分别设计了学术、生活、工作三个场景,在场景之下引导读者做任务,继而在任务的基础上锻炼读者主动解决问题和团队合作的能力。本教材的侧重点是英语知识,而不是对物联网技术知识的介绍。因为物联网专业的教师会通过更具体详细的课程来系统地教授物联网知识,所以在本教材的三章中仅有一章是介绍物联网基础知识的。读者在使用本教材时应已具备相关专业知识,通过阅读本教材的文章来了解专业中的基本知识如何用英语来表达,并学习英文术语。教材的第2章和第3章分别介绍了物联网的应用和政策考量,这种编排使全书更具人文特色,鼓励读者在学习技术的同时也要对技术所带来的社会生活文化甚至是伦理问题的变化进行思考。

入选本教材的15篇文章皆经过精心挑选,来自物联网技术的权威原版教材、影响力大的学术论文、标准制定组织发布的报告及语言和内容皆优质的英美期刊。编写团队在筛选和改编文章时充分考虑了读者的需求,控制文章的难度,并使其整体结构更合理,逻辑更严密。本教材的读者对象是普通高等院校物联网及相关专业的学生,同时本教材也适合人文学科专业对物联网技术感兴趣的学生以及职场人士使用。

在编写本教材之时,市场上已有多本"物联网英语"的相关教材,编者也做了详细的调研,阅读了大部分市场上能找到的教材。已有教材大多以文章阅读为主,并在课文后附上少量名词解释或者翻译习题。本教材相对于已有教材有着明显的优势。第一,有着专业而详细的语言技能讲解,如对逻辑地图、SQ3R(survey, question, read, recall and review)等阅读技能的讲解，对学术论文、说服性文章等写作技能的讲解。本教材能够授人以渔，提高读者的语言使用能力，而不仅仅是专业内容的灌输。第二，不同于市面上"物联网英语"相关方面的教材仅呈现文章和少量词汇翻译习题的做法，本教材涉及听、说、读、写全方位的训练，能够让读者灵活使用英语讨论与物联网相关的话题。第三，输入输出并重，或者以输出带动输入，课后设计了大量活动以方便教师在课堂上开展。第四，将术语和通用词汇分开，行业英语很重要的一部分就是术语的处理和学习，术语需要和通用词汇分开学习。第五，课文析出词汇使用英文进行词义的标注。这些析出重点词汇为大学英语四级及以上词汇，有利于降低阅读难度，读者若遇到生词可以及时在课文一侧找到释义。之所以没有像许多教材那样直接使用中文释义，而是放上英文释义，是想加强读者查阅英英词典的意识，或者使读者习惯使用英文来理解生词，这样读者可以更加准确地理解词汇以及词汇的用法。

本教材提供详细的语言技能讲解以及精心设计、相关性强的习题，读者在阅读文章后，首先可以通过课后阅读题来检验自己对文章的理解情况，其次可以根据阅读技巧的讲解来训练和提高自身的阅读能力。词汇和听力部分的练习也适合读者自学。

本教材的编排适合教师在课堂上开展活动。在每个章节的各个技能点中，尤其是口语和写作部分，编者设计了丰富多彩的场景化活动，将语言技能训练和物联网工作、生活场景完美结合。例如，第1单元的口语活动部分要求学生假设自己是一名教授物联网技术的教师，要去给中小学生做一个讲座，讲一些基本的物联网概念和应用。学生在准备这个讲座时需要充分考虑中小学生这类听众，因此准备的内容应简单易懂、图文并茂、趣味性强。在这种"产品"的导向下，学生不仅需要掌握物联网基本技术，而且需要练习如何用简单的英文对其进行表达。在这个过程中，学生的语言组织能力、信息搜索能力、利用专业知识解决问题的能力都能得到提升。

本教材的编写团队由北京邮电大学人文学院从事一线教学的5位青年教师组成，编者有着语言学、翻译学和外语教育的科研背景，并对信息科学技术知识抱有极大的热情。本教材凝结着编者在若干年教学中总结的精华。从最初的理念设计到选文、编写和校对的各个过程，编写团队出色的团队合作意识使整个教材的编写过程愉快且富有成效。本教材第1~3单元由许可老师编写，第4、5、9单元由任春蕾老师编写，第6~8单元由赵欣老师编写，第10、14、15单元由王姗姗老师编写，第11~13单元由刘佳老师编写。同时感谢北京邮电大学人文学院英语教研组的白天惠老师在教材初期的设计和选文上给予的大力支持，感谢"物联网工程专业教材丛书"总主编张锦南老师对编者团队的信任，感谢北京邮电大学出版社的姚顺编辑对本书出版的支持。

由于时间仓促和编写团队的能力有限，书中难免存在疏漏和错误，欢迎读者积极与编写团队沟通在教材使用和阅读过程中遇到的问题，以便我们能够向读者呈现更加优质的教材。

许　可

于北京邮电大学

目 录

Chapter 1　IoT Technology Fundamentals　　3

　Unit 1　From the Internet of Computers to the Internet of Things　　3

　Unit 2　Tagging Things—RFID　　14

　Unit 3　Wireless Sensor Networks　　23

　Unit 4　Machine-to-Machine　　40

　Unit 5　Setting the Standard　　54

Chapter 2　The Applications of the IoT　　69

　Unit 6　How the Internet of Things Can Prepare Cities for Natural Disasters　　69

　Unit 7　IoT in Action in Healthcare　　82

　Unit 8　Flying Smarter — The Smart Airport and the Internet of Things　　93

　Unit 9　Industrial Internet of Things　　104

　Unit 10　The IoT Data Opportunity for Logistics Companies Is here　　116

Chapter 3　IoT Policy Considerations　　131

　Unit 11　Social-ethical Considerations　　131

　Unit 12　The Implications of the Internet of Things on Social Media and Branding　　143

Unit 13　Security and the Internet of Things ··· 153

Unit 14　The Internet of Things Is Revolutionizing Our Lives，but Standards

　　　　Are a Must ·· 164

Unit 15　What the Internet of Things Means for Consumer Privacy ···················· 176

Word List ··· 188

Terminology List ·· 200

IoT Technology Fundamentals

- From the Internet of Computers to the Internet of Things
- Tagging Things – RFID
- Wireless Sensor Networks
- Machine-to-Machine
- Setting the Standard

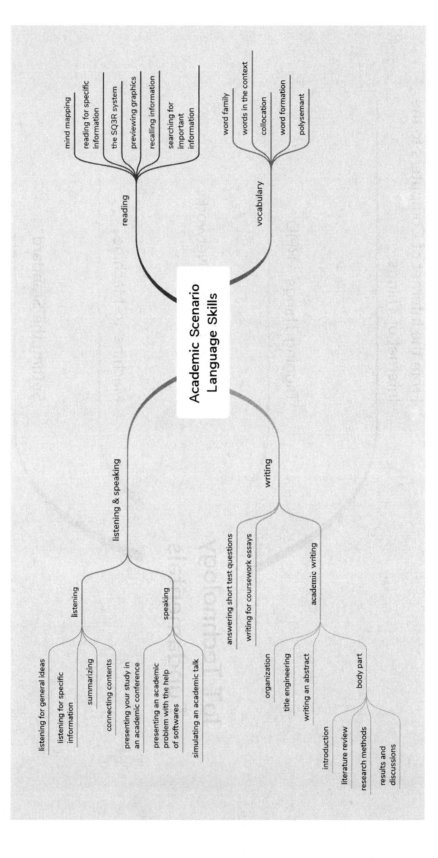

Chapter 1 IoT Technology Fundamentals

This Chapter provides 5 articles on some basic IoT technologies. You will be put in an academic scenario, in which you will learn how to read for academic purposes, write research papers, and communicate in academic conferences.

Unit 1   From the Internet of Computers to the Internet of Things

1   The Internet of Things represents a <u>vision</u> in which the Internet extends into the real world <u>embracing</u> everyday objects. Physical items are no longer disconnected from the virtual world, but can be controlled remotely and can act as physical access points to Internet services. An Internet of Things makes computing truly <u>ubiquitous</u>—a concept initially put forward by Mark Weiser in the early 1990s. This development is opening up huge opportunities for both the economy and individuals. However, it also involves risks and undoubtedly represents an <u>immense</u> technical and social challenge.

2   The Internet of Things vision <u>is grounded in</u> the belief that the steady advances in **microelectronics**, communications and information technology we have witnessed in recent years will continue into the <u>foreseeable</u> future. In fact—due to their <u>diminishing</u> size, constantly falling price and declining energy consumption—**processors, communications modules** and other electronic components are being increasingly <u>integrated</u> into everyday objects today.

3   "Smart" objects play a key role in the Internet of Things vision, since <u>embedded</u> communication and information technology would have the potential to <u>revolutionize</u> the use of these objects. Using sensors, they are able to <u>perceive</u> their

---

vision /ˈvɪʒən/ n.
a dream or similar experience, especially of a religious kind
embrace /ɪmˈbreɪs/ v.
to accept a set of beliefs, especially when it is done with enthusiasm
ubiquitous /juːˈbɪkwɪtəs/ adj.
seeming to be everywhere; very common
immense /ɪˈmɛns/ adj.
extremely large or great
be grounded in/on sth (to be) based on sth
foreseeable /fɔrˈsiəbəl/ adj.
that you can predict will happen
diminish /dɪˈmɪnɪʃ/ v.
to become or to make sth become smaller, weaker, etc.
diminishing adj.
integrate /ˈɪntɪˌɡreɪt/ v.
to combine two or more things so that they work together
embed /ɪmˈbɛd/ v.
to fix sth firmly into a substance or solid object
revolutionize /ˌrɛvəˈluːʃəˌnaɪz/ v.
to completely change the way that sth is done
perceive /pəˈsiːv/ v.
to notice

context, and via built-in networking capabilities they would be able to communicate with each other, access Internet services and interact with people. "Digitally upgrading" conventional object in this way enhances their physical function by adding the capabilities of digital objects, thus generating substantial added value. Forerunners of this development are already apparent today—more and more devices such as sewing machines, exercise bikes, electric toothbrushes, washing machines, electricity and photocopiers are being "computerized" and equipped with network interfaces.

4 In other application domains, Internet connectivity of everyday objects can be used to remotely determine their state so that information systems can collect up-to-date information on physical objects and processes. This enables many aspects of the real world to be "observed" at a previously unattained level of detail and at negligible cost. This would not only allow for a better understanding of the underlying processes, but also for more efficient control and management. The ability to react to events in the physical world in an automatic, rapid and informed manner not only opens up new opportunities for dealing with complex or critical situations, but also enables a wide variety of business processes to be optimized. The real-time interpretation of data from the physical world will most likely lead to the introduction of various novel business services and may deliver substantial economic and social benefits.

5 The use of the word "Internet" in the catchy term "Internet of Things" which stands for the vision outlined above can be seen as either simply a metaphor—in the same way that people use the Web today, things will soon also communicate with each other, use services, provide data and thus generate added value—or it can be interpreted in a stricter technical sense, postulating that an **IP protocol stack** will be used by smart things (or at least by the **"proxies"**, their representatives on the network).

6 The term "Internet of Things" was popularized by the work of the Auto-ID Center at the Massachusetts Institute of Technology (MIT), which in 1999 started to design and propagate a cross-company **RFID infrastructure**. In 2002, its co-founder and former head Kevin Ashton was quoted in *Forbes* Magazine as saying, "We need an internet for things, a standardized way for computers to understand the real world".

---

context /ˈkɒntɛkst/ n.
the situation in which sth happens
generate /ˈdʒɛnəˌreɪt/ v.
to produce or create sth
substantial /səbˈstænʃəl/ adj.
large in amount, value or importance
forerunner /ˈfɔːˌrʌnə/ n.
a person or thing that came before and influenced sb/sth else that is similar
attain /əˈteɪn/ v.
to succeed in getting sth, usually after a lot of effort
unattained adj.
negligible /ˈnɛglɪdʒəbəl/ adj.
of very little importance or size and not worth considering
efficient /ɪˈfɪʃənt/ adj.
doing sth well and thoroughly with no waste of time, money, or energy
in a ... manner the way that sth is done or happens
optimize /ˈɒptɪˌmaɪz/ v.
to make sth as good as it can be

catchy /ˈkætʃɪ/ adj.
pleasing and easily remembered
metaphor /ˈmɛtəfə/ n.
a way of describing something by referring to it as something different and suggesting that it has similar qualities to that thing
interpret /ɪnˈtɜːprɪt/ v.
to explain the meaning of sth
postulate /ˈpɒstjʊˌleɪt/ v.
to suggest or accept that sth is true so that it can be used as the basis for a theory, etc.

propagate /ˈprɒpəˌgeɪt/ v.
to spread an idea, a belief or a piece of information among many people

This article was entitled "The Internet of Things", and was the first documented use of the term in a literal sense. However, already in 1999 essentially the same notion was used by Neil Gershenfeld from the MIT Media Lab in his popular book *When Things Start to Think* when he wrote "in retrospect it looks like the rapid growth of the World Wide Web may have been just the trigger charge that is now setting off the real explosion, as things start to use the Net".

7   In recent years, the term "Internet of Things" has spread rapidly—in 2005 it could already be found in book titles, and in 2008 the first scientific conference was held in this research area. European politicians initially only used the term in the context of RFID technology, but the titles of the RFID conferences "From RFID to the Internet of Things" (2006) and "RFID: Towards the Internet of Things" (2007) held by the EU Commission already allude to a broader interpretation. Finally, in 2009, a dedicated EU Commission action plan ultimately saw the Internet of Things as a general evolution of the Internet "from a network of interconnected computers to a network of interconnected objects" (Internet of Things—An action plan for Europe, 2009).

8   From a technical point of view, the Internet of Things is not the result of a single novel technology; instead, several complementary technical developments provide capabilities that, taken together, help to bridge the gap between the virtual and physical world. These capabilities include communication and cooperation, **addressability**, identification, sensing, **actuation**, embedded information processing, **localization** and **user interfaces**.

Word count: 862

Source: Adapted from an updated translation of Mattern F, Floerkemeier C. Vom Internet der Computer zum Internet der Dinge[J]. Informatik-Spektrum, 33(2): 107-121.

---

in retrospect thinking about a past event or situation, often with a different opinion of it from the one you had at the time

trigger /ˈtrɪgə/ n.
the part of a bomb that causes it to explode

explosion /ɪkˈspləʊʒən/ n.
a large, sudden or rapid increase in the amount or number of sth

allude to sb/sth to mention sth in an indirect way

dedicated /ˈdedɪkeɪtɪd/ adj.
designed to do only one particular type of work

ultimately /ˈʌltɪmɪtli/ adv.
in the end; finally

evolution /ˌiːvəˈluːʃən/ n.
the gradual development of sth

complementary /ˌkɒmplɪˈmentəri/ adj.
things go well together, although they are usually different

virtual /ˈvɜːtʃuəl/ adj.
made to appear to exist by the use of computer software, for example on the Internet

physical /ˈfɪzɪkəl/ adj.
connected with things that actually exist or are present and can be seen, felt, etc. rather than things that only exist in a person's mind

I Reading

Reading for academic purposes

If you are going to read for academic purposes, you will find that most of the materials will be intended for "serious" readers. You will be exposed to language which is directly relevant to your specific fields of study. It is important to keep the reading purpose in mind while reading a text. This way, you can make decisions about the most effective way to read that text. For example, you can decide whether to read quickly through a text, or skip some of the materials and spend time reading carefully through the parts that help achieve the task.

When you read for academic purposes, you are probably searching for a topic to write about, finding evidences to support your ideas, updating your knowledge on the latest discoveries. Whatever your purpose may be, you need to acquire a general understanding. Creating a mind map may help you achieve that goal.

1. Reading and mind mapping

When you read, creating a mind map may help you understand the structure of the text, and figure out the main idea and supporting details. Glossing over part headings and the first sentences of each paragraph may help.

Task 1

Directions: Please complete the following graph, and fill in as many details as possible.

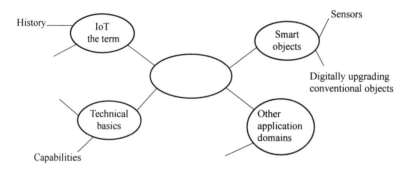

2. Reading for specific details

Task 2

Directions: Please answer the following questions.

(1) What enables the processors, communications modules and other electronic components to integrate into everyday objects today?

(2) Please list some of the conventional objects that have been "computerized" or "digitally upgraded" to generate added value.

(3) Which organization popularized the term "Internet of Things"?

(4) Apart from novel technologies, what are the complementary technical developments that enable the Internet of Things?

II Vocabulary

Key words and expressions

vision	embrace	ubiquitous	immense
be grounded in/on sth	foreseeable	diminish	integrate
embed	revolutionize	perceive	context
generate	substantial	forerunner	attain
negligible	efficient	in a… manner	optimize
catchy	metaphor	interpret	postulate
propagate	in retrospect	trigger	explosion
allude to sb/sth	dedicated	ultimately	evolution
complementary	virtual	physical	

Terminology

microelectronics	n.	微电子学
processor	n.	处理机,处理器
communication module	n.	通信模块
IP protocol stack	n.	互联网通信栈
proxy	n.	代理
RFID (Radio Frequency Identification infrastructure)	n.	无线射频识别基础设施
addressability	n.	寻址能力
actuation	n.	驱动
localization	n.	本地化
user interface (UI)	n.	用户界面

1. Word bank

Task 1

Directions: Please fill in the blanks with the words listed below. You may need to change the form when necessary.

ubiquitous	context	diminish	optimize	interpret
physical	evolution	explosion	virtual	postulate
substantial	efficient	perceive	integrate	generate

(1) Science fiction stories and movies can also be used to deal with ethical（道德的）issues. Fictional examples can often be more effective than historical or current events, because they explore ethical issues in a _____ that students often find interesting and that is independent of current political or economic considerations.

(2) Researcher has found that earnings are higher when the academic major is _____ as closely related to employment.

(3) Susan was assigned a difficult task. She couldn't sleep well for many days and felt that she might one day _____ under stress.

(4) The _____ of management training at McDonald's has clearly reached a new level, as a new generation of managers can create and sustain successful business environments.

(5) This charity was trying to refurbish an existing building in the village but required _____ funding to bring it up to a suitable standard.

(6) What was once the wild dream of spies—to plant a "bug" in every object—has been enlarged and re-shaped into the millennial dream of _____ computing.

(7) Our current management system is too old, and thus should be _____.

(8) Throughout his political career, he has taken _____ every position on every issue.

(9) Steve Wozniak had a genius for designing the most _____ computer from the least number of parts.

(10) These sensors will _____ enormous amount of data we never had before from the physical world.

2. Word family

One way to figure out the meaning of an unknown word is to look for its relationship with other words in the same word family. Even if you cannot figure out the exact meaning, your understanding can be enough to allow you to read on.

Task 2

Directions: Please look at the phrases from the text. Write down at least one other word you know that is related to the underlined word. An example has been given to you.

(1) the foreseeable future   foresee

(2) revolutionize the use of these objects   _____

(3) at negligible cost
(4) a dedicated EU Commission action plan
(5) complementary technical developments
(6) propagate a cross-company RFID infrastructure

Ⅲ Listening and Speaking

Directions: Scan the QR code on the margin and you will watch a video clip called "How It Works: The Internet of Things created by IBM Think Academy".

Video Clip

Listening preparation—vocabulary

Please study the following new words that will appear in the listening material.

sophisticated	*adj.*	(of a machine, system, etc.) clever and complicated in the way that it works or is presented
analytics	*n. pl.*	analyses
asset	*n.*	a thing of value, especially property, that a person or company owns, which can be used or sold to pay debts
coupon	*n.*	a small piece of printed paper that you can exchange for sth or that gives you the right to buy sth at a cheaper price than normal
warranty	*n.*	a written agreement in which a company selling sth promises to repair or replace it if there is a problem within a particular period of time
pinpoint	*v.*	to be able to give exact reason for sth or to describe sth exactly

Task 1 Spot dictation

Directions: Please watch the first part (1 minute) of the video, and fill in the blanks below. You may watch the video several times to get the correct words.

The Internet of Things is changing much of the world we live in, from the way we drive and how we make (1)_____, and even how we get energy for our homes. Sophisticated (2)_____ and chips are embedded in the (3)_____ things that surround us, each transmitting valuable data, data that lets us better understand how these things work and work together. But how exactly do all these devices share such large (4)_____ of data, and how do we put that information to work? Whether we're improving the production of a factory, giving city residents real time (5)_____ on where to park or monitoring our personal health. It's the common Internet of Things platform that brings us (6)_____ information together and provides the common languages for the devices and apps to (7)_____ with each other. The process starts with the devices themselves which securely communicate with the Internet of Things platform. This platform (8)_____ data from many devices and applies analytics to share the most valuable data with applications that address industry specific needs. Let's start with a simple example, a car.

Task 2　Summarizing

Directions: please watch the second part, take notes and summarize, using the car as an example of IoT application. Here are some questions that may help you to summarize. You may present it to your classmates.

(1) What happened to Rebecca's car?

(2) What is a diagnostic bus in Rebecca's car? What does it do?

(3) How does the manufacturer gather information about thousands of cars?

(4) What do manufacturers do with all the information gathered?

Task 3　Public speaking: a lecture

Directions: Suppose you are going to give a lecture on the basics of IoT for middle school students. You should try your best to make it easy to understand, interesting and inspiring. You may use the reading material in this chapter as a point of reference and add your own examples on the application scenarios.

Ⅳ　Writing

Writing for coursework and exam

Answering short-answer test question

You will often be asked to write short answers to questions about texts you read in college. It is a good idea to use words from the question in your answer, to use only relevant information from the text, and to give examples where possible. Copying the original text is never a good idea for these kinds of questions. You should paraphrase.

The best way to prepare for short-answer questions is to make up some questions by yourself and try to answer them.

Task 1　Answer the question

Directions: According to the text, what contributes to the evolution of the Internet of Things paradigm, in which smart objects actually communicate with each other?

_____
_____
_____
_____

Task 2

Directions: Read the text, try to make up some questions by yourself, and write them down on the lines below. Then work in pairs, and answer each other's questions.

_____
_____
_____
_____

KEY

I Reading

Task 1

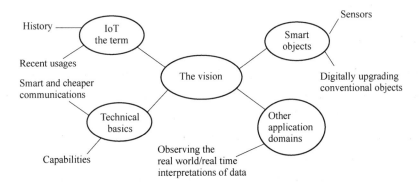

Task 2

(1) diminishing size, constantly falling price, declining energy consumption

(2) sewing machines, exercise bikes, electric toothbrushes, washing machines, electricity meters, photocopiers

(3) the Auto-ID Center at the Massachusetts Institute of Technology (MIT)

(4) communication and cooperation, addressability, identification, sensing, actuation, embedded information processing, localization and user interfaces

II Vocabulary

Task 1

(1) context (2) perceived (3) explode (4) evolution (5) substantial (6) ubiquitous (7) optimized (8) virtually (9) efficient (10) generate

Task 2

(2) revolution (3) neglect, negligence, negligent (4) dedicate, dedication (5) complement, complemental, complementarity (6) propagation propaganda

Ⅲ Listening and Speaking

Task 1

(1) purchases (2) sensors (3) physical (4) quantities (5) updates (6) diverse (7) communicate (8) integrates

Ⅳ Writing

Task 1

Many key technologies have enabled smart objects to communicate with each other, such as smaller wireless communications modules, the application of IPv6, flash memory equipped with high capacity, processor with lower energy requirement pre-instruction, smart phones equipped with bar code and NFC, etc.

Script for Listening

The Internet of Things is changing much about the world we live in, from the way we drive to how we make purchases, and even how we get energy for our homes. Sophisticated sensors and chips are embedded in the physical things that surround us, each transmitting valuable data, data that lets us better understand how these things work and work together. But how exactly do all these devices share such large quantities of data, and how do we put that information to work? Whether we're improving the production of a factory, giving city residents real time updates on where to park or monitoring our personal health. It's the common Internet of Things platform that brings us diverse information together and provides the common language for the devices and apps to communicate with each other. The process starts with the devices themselves which securely communicate with an Internet of Things platform. This platform integrates the data from many devices and applies analytics to share the most valuable data with applications that address industry specific needs.

Let's start with a simple example—a car. After taking a long road trip, Rebecca notices that her check engine light is come on [SIC]. She knows that she needs to have her car looked at by a mechanic, but is not sure whether it's something minor or something that needs immediate attention. As it turns out, the sensor that triggered Rebecca check engine light monitors the pressure in a brake line. This sensor is one of many monitoring processes throughout the car, which are constantly communicating with each other.

A component in the car called the diagnostic bus gathers the data from all these sensors then passes it to a gateway in the car. The gateway integrates and sorts the data from the sensors. This way, only the most relevant diagnostic information will be transmitted to the manufacturers' platform. But before sending this organized data, the cars gateway and platform must first register with each other and confirm a secure communication. The platform is constantly gathering and storing thousands of bits of information from Rebecca's car and hundreds of thousands of cars like hers, building a historical record and a secure database. The manufacturers added rules and logic to the platform.

So when Rebecca's car senses signal that her brake fluid has dropped below a recommended level, the platform triggers an alert in her car. The manufacturer also uses the platform to create and manage applications that solve specific issues. In this case the manufacturer can deploy an application on the platform called the asset management system. This application oversees all of their customers cars on the road as well as all the parts in their warehouses. It uses the data from Rebecca's car to offer her a potential appointment time to service her car, directions to the nearest certified dealer, and a coupon for the service.

What's more, the Apple ensures that Rebecca's breaks are covered under her warranty that the correct replacement part is ordered and then sent to the dealership. So it is ready when she arrives. But the manufacturers' analysis does not stop there. They have also deployed a continuous engineering application that tracks not only Rebecca's car but hundreds of thousands of others, looking for ways to improve the design and manufacturing process of the car itself. If the same problem in her brake line crops up in a critical number of other cars, the manufacturer uses applications custom built for the automobile industry to pinpoint the exact problem.

They can see if these cars were made at the same factory, used the same parts or came off the assembly line on the same day. So what do all these pieces add up to? Streamlined inventory management for the dealer, a better safer car from the manufacturer, and for Rebecca? It means she can be back on the road faster and get to where she's going safely. All thanks to the Internet of Things.

Unit 2   Tagging Things—RFID

1   For information and communication access to be truly and seamlessly embedded in the environment surrounding us, the exponential growth of networked devices must be accompanied by a paradigm shift in computing. Such a paradigm shift will mean that smart computers will become a common item in many households.

2   Delivering on the promise of the Internet of Things is, however, currently limited by our inability to collect raw data about things, their location and status. Radio Frequency Identification (RFID) provides just such a capability and is a key enabler of a ubiquitous communication environment. RFID refers to those technologies that use radio waves to automatically identify and track individual items. In this respect, it can be conceptualized as analogous to common short-range wireless technologies such as **ZigBee**, but with much higher computing and tracking capabilities.

3   RFID is not new, as it is based on radio, which dates back to the early understanding of **electromagnetic energy** by Michael Faraday from the 1840s and was expanded into popular use in the early 20th century. Shortly after, the 1920s saw the birth of radar, which detects and locates objects (their position and speed) through the reflection of radio waves. RFID combines radio technology with radar and dates back to the seminal 1948 paper by Harry Stockman "Communication by Means of Reflected Power". Although RFID is not new, mass-market applications have only been developed over the last decade.

Technical overview of RFID

4   Technically speaking, RFID systems consist of three main components (Figure 2-1):
- A **transponder** or tag to carry data, which is located on the object to be identified. This normally consists of a coupling element (such as a **coil**, or **microwave antenna**) and an electronic microchip, less than 1/3 millimetre in size. Tags can be passive, semi-passive or active, based on their power source and the way they are used, and can be read-only, read/write

---

seamless /ˈsiːmlɪs/ *adj*.
with no spaces or pauses between one part and the next
seamlessly *adv*.
exponential /ˌekspəʊˈnenʃəl/ *adj*.
(of a rate of increase) becoming faster and faster
raw /rɔː/ *adj*.
not yet organized into a form in which it can be easily used or understood
conceptualize /kənˈseptjʊəˌlaɪz/ *v*.
to form an idea of sth in your mind
analogous /əˈnæləɡəs/ *adj*.
similar in some way to another thing or situation and therefore able to be compared with it

seminal /ˈsemɪnəl/ *adj*.
(formal) very important and having a strong influence on later developments

passive /ˈpæsɪv/ *adj*.
accepting what happens or what people do without trying to change anything or oppose them

or read/write/re-write, depending on how their data is encoded. Tags do not need an in-built power source, as they take the energy they need from the electro-magnetic field <u>emitted</u> by readers.

emit /ɪˈmɪt/ v.
(formal) to send out sth such as light, heat, sound, gas, etc.

- An **interrogator** or reader, which reads the transmitted data (e.g. on a device that is handheld or embedded in a wall). Compared with tags, readers are larger, more expensive and power-hungry.
- **Middleware** (such as RS232, RS485 etc.), which <u>forwards</u> the data to another system, such as a database, a personal computer or a robot control system.

forward /ˈfɔːwəd/ v.
(formal) to send or pass goods or information to sb

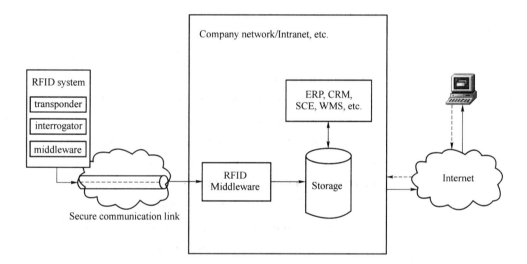

Figure 2-1 An RFID system in a network architecture

5 In the most common type of system, the reader transmits a low-power radio signal to power the tag (which, like the reader, has its own antenna). The tag then selectively reflects energy and thus transmits some data back to the reader, communicating its identity, location and any other relevant information. Most tags are passive, and activated only when they are within the coverage area of the interrogator. While outside this area, they remain <u>dormant</u>. Information on the tag can be received and read by readers and then forwarded to a computer database. Frequencies currently used for data transmission by RFID typically include 125 **kHz** (low **frequency**), 13.56 MHz (high frequency), or 800~960 MHz (ultra high frequency). RFID standards relate both to frequency protocols (for data communication) and data

dormant /ˈdɔːmənt/ adj.
not active or growing now but able to become active or to grow in the future

format (for data storage on the tag). Depending on their construction, RFID tags can be less than a square millimetre in area and thinner than a sheet of paper. One of the most <u>pivotal</u> aspects of these electronic labels is that they allow for the <u>accurate</u> identification of objects and the forwarding of this information to a database stored on the internet or on a remote server. In this manner, data and information processing capabilities can be associated with any kind of object. This means that not only people, but also things will become connected and contactable. In the most common application of RFID, for supply chain management, it is typically used as a long-term <u>enhancement</u> of the traditional bar code. But RFID tags represent much more than the next generation of bar codes and have many unique advantages. Traditional bar codes identify only a category of product. For instance, all Gillette Mach 3 razor <u>blades</u> have the same bar code. However, with RFID tags, each pack of blades would have its own unique **identifier** that can be transmitted to suitably located readers for monitoring. The RFID tag can hold much more data than a bar code, and becomes in some sense a mini-database embedded in the item. Currently, the **Electronic Product Code (EPC)** is the <u>dominant</u> standard for data contained in RFID tags for the purpose of item-level tracking.

6 RFID also allows data <u>capture</u> without the need for a line of sight ("RFID: Powering the supply chain", 2004). Some applications limit the read range of RFID tags to between 0.15～0.20 meters, but the majority have a range of around a meter. Newer tags in the UHF frequency bands could even have a range of 6～7.5 meters. This means that physical **manipulation** or access to individual items (often stacked or piled) is not needed for identification and tracking. This is not the case with the bar code, which must be "seen" at close range by scanners in order to be identified. Depending on whether tags are read-only, read/write or read/write/re-write, tags can create a variety of interfaces that can connect computers directly to individual physical items, and even to people, thus promising a truly ubiquitous future.

Word count: 920

Source: Adapted from ITU. The Internet of Things[R]. Geneva: ITU, 2005.

---

pivotal /ˈpɪvətəl/ *adj.*
of great importance because other things depend on it

accurate /ˈækjərɪt/ *adj.*
able to give completely correct information or to do sth in an exact way

enhance /ɪnˈhɑːns/ *v.*
to increase or further improve the good quality, value or status of sb/sth
enhancement *n.*

blade /bleɪd/ *n.*
the flat part of a knife, tool or machine, which has a sharp edge or edges for cutting

dominant /ˈdɒmɪnənt/ *adj.*
more important, powerful or noticeable than other things

capture /ˈkæptʃə/ *n.*
the act of putting sth into a computer in a form it can use

I Reading

The SQ3R system

SQ3R means survey (S), question (Q), read, recall and review (3R). These five steps may help you to become an active reader. They are also especially useful in reading academic texts.

First step: Survey

When surveying a text, you should look at the title, section, headings, tables, graphs and images. If you are reading a piece of academic writing, you should also scan its abstract, key words (or index terms) and references. Surveying saves your time, since you may decide whether the article is worth reading.

Task 1

Direction: Please survey "Tagging Things—RFID" and report to the class what you have found.

Second step: Question

After surveying the text, you should begin to question. This step may help you think, instead of passively reading the text directly. Here are some questions you may ask yourself: "What do I already know about this?" "What more do I need to know?" With the questions in mind, you can check the answers during reading.

Task 2

Directions: Before you read, please answer the following questions.
1. Have you ever heard of RFID? Please write down your current understanding of RFID.
_____
_____

2. Please write down your own questions after surveying the text.
_____
_____

Third step: Read

In this step, you may skim the text for a general idea and scan it for specific details.

Task 3

*Directions: Please decide whether the following statements are **true**, **false** or **not given** according to the text.*
1. RFID has lower computing and tracking capabilities compared to ZigBee.

2. RFID was invented before radar.
3. Tags are power hungry. They need batteries.
4. Readers are larger and more expensive than tags.
5. RFID is used to enhance industrial efficiency.
6. Bar codes can only identify a category of product, but RFID tags give each item a unique identifier.

Fourth step: Recall

To recall means to "remember". You can do this at various stages in reading a text—for example, in the case of longer texts, after reading paragraphs and/or sections, you should cover up the text you have read and try to remember the main point.

Task 4

Directions: Here is a table adapted from the text showing a contrast between traditional bar code and RFID. Please do not refer to the text, and try your best to fill in the blanks in the table.

Traditional bar codes	RFID
	Unique identification of individual items, allowing databases of specific item/location information to be generated, giving each item its own identity for real-time identification and tracking
	Data capture without the need for line of sight or physical manipulation
Data can be saved only once	
Hold less data	
Less potential to intrude on privacy	Privacy-enhancing technologies can be used to kill or block tags

Fifth step: Review

At this stage you should check the text again to make sure your notes are accurate and you have written down everything you need for the text.

Task 5

Directions: Now you can go back to read the text and check your answers of the table in Task 4.

II Vocabulary

Key words and expressions

seamless	exponential	raw	conceptualize
analogous	seminal	emit	forward
dormant	pivotal	accurate	enhance
blade	dominant	capture	passive

Chapter 1 IoT Technology Fundamentals

Terminology

ZigBee	n.	紫蜂
electromagnetic energy	n.	电磁能
transponder	n.	发射机应答器,询问机,转发器
coil	n.	线圈
microwave antenna	n.	微波天线
interrogator	n.	询问机
middleware	n.	中间件
kilohertz(kHz)	n.	千赫
frequency	n.	频率
identifier	n.	标识符
Electronic Product Code (EPC)	n.	电子产品码
manipulation	n.	操作,操控

Understanding words without referring to the dictionary

There are several ways you can understand new words without referring to the dictionary. You may guess the meaning from the context or find definition for words. In order to help their readers, many authors explain or define new terms in their writing.

The language cues may be *or*, *refers to*, *means that*, *is defined as*, *can be explained as* …

An example is the word "transponder". You may find its explanation directly following the word.

> A transponder or tag to carry data, which is located on the object to be identified. This normally consists of a coupling element (such as a coil, or microwave antenna) and an electronic microchip, less than 1/3 millimeter in size.

Task 1

Directions: Now please locate the definition or explanation for the following words in the text.

1. RFID

___

2. radar

___

3. interrogator

___

Task 2 Word bank

Directions: Please fill in the blanks with the words listed in the key words and expressions box above. You may need to change the form when necessary.

1. China is already at the front when it comes to blurring lines between social media, search and e-commerce, with social platforms like Tencent integrating _____ payment and shopping functions, all inside one walled garden.

2. I hope the start-up Primetime Tours will provide consumers with _____ information about all the housing options available to them. Understanding what the options are and the needs they fill is the first step to making an informed choice.

3. These photographs _____ the essence of working-class life at the turn of the century.

4. Barack Obama's election as America's first black president remains a _____ moment in the country's history.

5. Florian's anger appeared from nowhere, a _____ volcano raging into life.

6. As an architect, you _____, plan and develop designs for the construction and renovation of commercial, institutional and residential buildings.

7. Communication, computer, and organizational skills are _____ for this job.

8. Marine construction technology like this is very complex, somewhat _____ to trying to build a bridge under water.

9. Television watching can become a _____, deadening activity if adults do not guide their children's viewing and fail to teach children to watch critically and learn from what they watch.

10. Warm weather along with the right nutrients（营养物质） in lakes, such as phosphates（磷） from agricultural runoff, creates the perfect environment for the _____ growth of blue-green algae（水藻）, also known as harmful algae blooms.

Ⅲ Speaking

Task Presenting your study in an academic conference

Directions: Suppose you are a researcher on IoT. Your recent paper on RFID is accepted by a major academic conference to be held in California. According to the invitation, you will be given 10 minutes to present your study, and another 10 minutes for question and answer. Please select and download a research paper on RFID, and make a 10-minute presentation in (for the) class.

Ⅳ Writing

Writing for coursework essay—traditional 3 or 5 paragraphs essay

A 3-paragraph traditional essay includes an introductory paragraph, a body paragraph and a concluding paragraph. A 5-paragraph traditional essay expands the body part into 3 paragraphs, and keeps the introductory and concluding ones.

In order to write clearly and logically, you always need a central idea, or point to develop in an essay. It is called thesis statement, and should appear in the introductory paragraph. Then you need specific evidence, or support, which appears in the body paragraph to back up the thesis. Lastly, you may write summary and closing remarks in the concluding paragraph.

<u>Writing an outline</u>

Here is a diagram about an outline for a 5-paragraph essay.

Title of the essay

Introduction: opening remarks thesis statement

Body paragraph 1: argument 1/ topic sentence 1 supporting evidence

Body paragraph 2: argument 2/ topic sentence 2 supporting evidence

Body paragraph 3: argument 3/ topic sentence 3 supporting evidence

Concluding paragraph: summary closing remark

Task

Directions: Please write an outline for a 5-paragraph essay on the application of RFID.

Title: The application of RFID

P1: introduction　　　　　_____

P2: topic sentence 1　　_____

supporting evidence 1　_____

P3: topic sentence 2　　_____

supporting evidence 2
P4: topic sentence 3
supporting evidence 3
P5: conclusion

KEY

Ⅰ Reading

Task 3

1. F 2. F 3. F 4. T 5. not given 6. T

Task 4

Traditional bar codes	RFID
Typically identify only a category of product	Unique identification of individual items, allowing databases of specific item/location information to be generated, giving each item its own identity for real-time identification and tracking
Require close-range scanning, typically with physical manipulation	Data capture without the need for line of sight or physical manipulation
Data can be saved only once	Tags can be passive or active, and also read-only, read/write or read/write/re-write
Hold less data	Hold more data, like a mini-database
Less potential to intrude on privacy	Privacy-enhancing technologies can be used to kill or block tags

Ⅱ Vocabulary

Task 1

1. RFID refers to those technologies that use radio waves to automatically identify and track individual items.

2. Radar detects and locates objects (their position and speed) through the reflection of radio waves.

3. An interrogator can also be called a reader. It forwards the data to another system, such as a database, a personal computer or a robot control system.

Task 2

1. seamless 2. accurate 3. capture 4. seminal 5. dormant 6. conceptualize

7. pivotal 8. analogous 9. passive 10. exponential

Unit 3　Wireless Sensor Networks

Characteristic features of WSNs

1　A WSN(Wireless Sensor Network) can generally be described as a network of **nodes** that cooperatively sense and control the environment, enabling interaction between persons or computers and the surrounding environment. WSNs nowadays usually include sensor nodes, actuator nodes, **gateways** and **clients**. A large number of sensor nodes **deployed** randomly inside of or near the monitoring area (**sensor field**), form networks through self-organization. Sensor nodes monitor the collected data to transmit along to other sensor nodes by **hopping**. During the process of transmission, monitored data may be handled by multiple nodes to get to gateway node after **multi-hop routing**, and finally reach the management node through the internet or satellite. It is the user who **configures** and manages the WSN with the management node, publishes monitoring missions and collects the monitored data.

2　As related technologies mature, the cost of WSN equipment has dropped dramatically, and their applications are gradually expanding from the military areas to industrial and commercial fields. Meanwhile, standards for WSN technology have been well developed, such as Zigbee®, WirelessHart, ISA 100. 11a, wireless networks for industrial automation-process automation (WIA-PA), etc. Moreover, with new application modes of WSN emerging in industrial automation and home applications, the total market size of WSN applications will continue to grow rapidly.

Sensor nodes

3　The sensor node is one of the main parts of a WSN. The hardware of a sensor node generally includes four parts: the power and power management module, a sensor, a microcontroller, and a wireless **transceiver** (Figure 3-1). The power module offers the reliable power needed for the system. The sensor is the bond of a WSN node which can obtain the

---

cooperative /kəʊˈɒpərətɪv/ *adj.*
involving doing sth together or working together with others towards a shared aim
cooperatively *adv.*
surrounding /səˈraʊndɪŋ/ *adj.*
that is near or around sth

random /ˈrændəm/ *adj.*
done, chosen, etc. without sb deciding in advance what is going to happen, or without any regular pattern
randomly *adv.*
dramatic /drəˈmætɪk/ *adj.*
(of a change, an event, etc.) sudden, very great and often surprising
dramatically *adv.*

emerge /ɪˈmɜːdʒ/ *v.*
to start to exist; to appear or become known

reliable/ rɪˈlaɪəbəl/ *adj.*
that can be trusted to do sth well; that you can rely on
bond /bɒnd/ *n.*
the way in which two things are joined together
obtain /əbˈteɪn/ *v.*
to get sth, especially by making an effort

environmental and equipment status. A sensor is in charge of collecting and transforming the signals, such as light, <u>vibration</u> and chemical signals, into electrical signals and then transferring them to the microcontroller. The microcontroller receives the data from the sensor and <u>processes</u> the data accordingly. The Wireless Transceiver (RF module) then transfers the data, so that the physical <u>realization</u> of communication can be achieved.

4 It is important that the design of the all parts of a WSN node consider the WSN node features of tiny size and limited power.

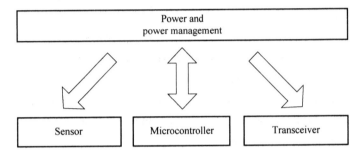

Figure 3-1 Hardware structure of WSN sensor node

Access network technologies

5 The access network, whose length ranges from a few hundred meters to several miles, includes all the devices between the **backbone network** and the **user terminals**. It is thus <u>aptly</u> called "the last mile". Because the backbone network usually uses **optical fiber** structure with a high transmission rate, the access network has become the <u>bottleneck</u> of the entire network system.

6 As shown in Figure 3-2, due to the open property of wireless channels, conflicts will happen in time, space or frequency <u>dimension</u> when the channel is shared among multiple users. The function of access network technologies is to manage and coordinate the use of channels resources to ensure the interconnection and communication of multiple users on the shared channel.

7 According to the distance and speed of access, existing access technologies can be <u>classified</u> into four categories: **wireless local area network (WLAN)**, **wireless metropolitan area network (WMAN)**, **wireless personal area network (WPAN)** and **wireless wide area network (WWAN)**. However, the overall developing trend of high transmission rates is not suitable for the application of WSNs, because it requires more reliable and stricter <u>real-time</u>

communication in a usually harsh environment, and low energy consumption to reduce maintenance cost.

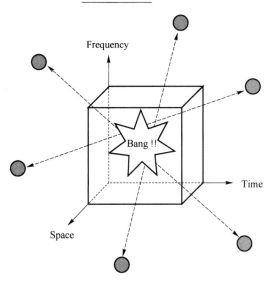

Figure 3-2   Access technologies

Topology

8   Generally, a WSN consists of a number of sensor network nodes and a gateway for the connection to the internet. The general deployment process of a WSN is as follows (Figure 3-3): firstly, the sensor network nodes **broadcast** their status to the surroundings and receive status from other nodes to detect each other. Secondly, the sensor network nodes are organized into a connected network according to a certain **topology** (**linear**, **star**, **tree**, **mesh**, etc.). Finally, suitable paths are computed on the constructed network for transmitting the sensing data. The power of sensor network nodes is usually provided by batteries, so the transmission distance of WSN nodes is short. The transmission distance can be up to 800 to 1,000 meters in the open outdoor environment with line of sight. It will sharply decline in the case of a sheltered indoor environment to an estimated few meters. In order to expand the coverage of a network, the sensor network uses multi-hop transmission mode. That is to say the sensor network nodes are both transmitter and receiver. The first sensor network node, the source node, sends data to a nearby node for data transmission to the gateway. The nearby node forwards the data to one of its nearby nodes that are on the path towards the gateway. The forwarding is repeated until the data arrives at the gateway, the destination.

maintenance /ˈmeɪntɪnəns/ n.
the act of keeping sth in good condition by checking or repairing it regularly

line of sight
an imaginary line that goes from sb's eye to sth that they are looking at
shelter /ˈʃɛltə/ v.
to stay in a place that protects you from the weather or from danger
sheltered adj.
estimate /ˈɛstɪmɪt/ v.
to form an idea of the cost, value, etc. of sth, but without calculating it exactly
coverage /ˈkʌvərɪdʒ/ n.
the amount of sth that sth provides; the amount or way that sth covers an area

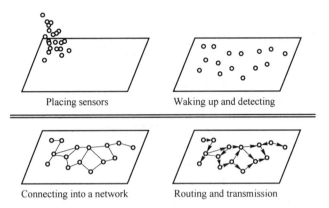

Figure 3-3　Organizing and transmitting process of WSNs

Data aggregation

9　In the energy-<u>constrained</u> sensor network environments, it is unsuitable in numerous aspects of battery power, processing ability, storage capacity and communication **bandwidth**, for each node to transmit data to the **sink node**. This is because in sensor networks with high coverage, the information reported by the neighboring nodes has some degree of **redundancy**, thus transmitting data separately in each node while consuming bandwidth and energy of the whole sensor network, which <u>shortens</u> lifetime of the network.

10　To avoid the above mentioned problems, data aggregation techniques have been introduced. Data aggregation is the process of <u>integrating</u> multiple copies of information into one copy, which is effective and able to meet user needs in middle sensor nodes.

11　The introduction of data aggregation benefits both from saving energy and obtaining accurate information. The energy consumed in transmitting data is much greater than that in processing data in sensor networks. Therefore, with the node's local computing and storage capacity, data aggregating operations are made to remove large <u>quantities</u> of redundant information, so as to <u>minimize</u> the amount of transmission and save energy. In the complex network environment, it is difficult to ensure the accuracy of the information obtained only by collecting few samples of data from the <u>distributed</u> sensor nodes. As a result, monitoring the data of the same object requires the <u>collaborative</u> work of multiple sensors which effectively improves the accuracy and the reliability of the information obtained.

---

constrain /kən'streɪn/ v.
to restrict or limit sb/sth

shorten /'ʃɔːtən/ v.
to make sth shorter; to become shorter

integrate /'ɪntɪˌɡreɪt/ v.
to combine two or more things so that they work together

quantity /'kwɒntɪtɪ/ n.
A quantity is an amount

minimize /'mɪnɪˌmaɪz/ v.
to reduce sth, especially sth bad, to the lowest possible level

distribute /dɪ'strɪbjuːt/ v.
to spread sth, or different parts of sth, over an area

collaborate /kə'læbəˌreɪt/ v.
to work together with sb in order to produce or achieve sth

collaborative adj.

12 The performance of data aggregation **protocol** is closely related to the network topology. It is then possible to analyze some data aggregation protocols according to star, tree, and chain network topologies as seen in Figure 3-4.

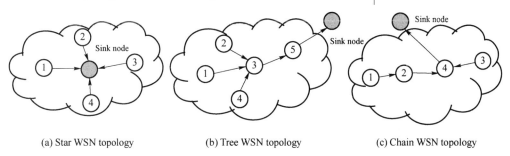

(a) Star WSN topology (b) Tree WSN topology (c) Chain WSN topology

Figure 3-4 Three kinds of WSN topologies: star, tree, chain

Word count: 1,130

Source: Adapted from International Electrotechnical Commission (IEC). Internet of Things: Wireless sensor networks[R]. Geneva: IEC, 2014.

I Reading

Preparing to read

1. Previewing graphics

Before reading a text, it is helpful to look at any graphs, tables, or diagrams. These will give you an idea of the content of the text.

Task 1

Directions: Look at the figures and read the words that explain them. Answer the questions below in your own words.

(1) What does Figure 3-1 show?

_____
_____

(2) According to Figure 3-2, what can be the problem caused by open properties of wireless channels?

_____
_____

(3) Please describe the organizing and transmitting process of WSNs illustrated in Figure 3-3.

_____
_____

After you read

2. Paraphrase

A paraphrase is a restatement of the meaning of a text or passage using other words. Paraphrasing is an excellent way to deepen your understanding and activate your language.

Task 2

Directions: Please paraphrase the following sentences picked from the text. Do not use the underlined words in your paraphrase. You are encouraged to change the sentence structure. You may refer to the explanations on the side of the text for help.

(1) The power module offers the <u>reliable</u> power needed for the system. The sensor is the <u>bond</u> of a WSN node which can <u>obtain</u> the environmental and equipment status.

_____
_____

(2) Because the backbone network usually uses optical fiber structure with a high transmission rate, the access network has become the <u>bottleneck</u> of the entire network system.

_____
_____

(3) In the energy-<u>constrained</u> sensor network environments, <u>it is unsuitable in numerous aspects</u> of battery power, processing ability, storage capacity and communication bandwidth, for each node to transmit data to the sink node.

_____
_____

3. Reading for specific information

Task 3

Directions: Complete the note below for a better understanding of key terms. One example is given.

WSN　　A WSN can generally be described as a network of nodes that cooperatively sense and control the environment, enabling interaction between persons or computers and the surrounding environment.

sensor nodes

the power module

access network technologies

multi-hop transmission mode

data aggregation

II Vocabulary

Key words and expressions

cooperative	surrounding	random	dramatic
emerge	reliable	bond	obtain
vibrate	process	realization	backbone
apt	bottleneck	dimension	classify
real-time	maintenance	line of sight	shelter
estimate	coverage	constrain	shorten
integrate	quantity	minimize	distribute
collaborate			

Collocation and usage

Collocation is the way words combine in a language to produce natural-sounding speech and writing. For example, in English you say strong wind but heavy rain. It would not be normal to say * *heavy wind* or * *strong rain*. To a native-speaker these combinations are highly predictable; to a learner they are anything but.

Task

Directions: Please look up the verbs in the box above in the dictionary for their collocations with prepositions, and fill in the blanks.

1. A number of significant points emerge _____ these studies.
2. White families have a biased image of African Americans as slum dwellers and who

are unable to integrate _____ American society.

3. Abby was a lively and out-going girl. Where she had been always vibrating _____ life.

4. Some scholars classify police violence _____ a public health problem.

5. Anglers（钓鱼者）are required to obtain prior authorization _____ the park keeper.

6. The apples are processed _____ juice or sauce.

7. She felt tightly constrained _____ her family commitments（责任）.

8. We have collaborated closely _____ the university on this project.

Terminology

wireless sensor networks（WSN）	n.	无线传感器网络
node	n.	节点
gateway	n.	网关
client	n.	客户端
deploy	v.	部署
sensor field	n.	传感域
hop	v.	跳跃
multi-hop routing	n.	多跳路由
configure	v.	配置
transceiver	n.	收发器
backbone network	n.	中枢网络
user terminal	n.	用户终端
optical fiber	n.	光纤
wireless local area network（WLAN）	n.	无线局域网
wireless metropolitan area network（WMAN）	n.	无线城域网
wireless personal area network（WPAN）	n.	无线个人网络
wireless wide area network（WWAN）	n.	无线广域网
broadcast	v.	广播
topology	n.	拓扑
linear	adj.	线性
star	adj.	星形
tree	adj.	树形
mesh	adj.	网状
aggregation	n.	聚合
bandwidth	n.	带宽
sink node	n.	汇点
redundancy	n.	冗余
protocol	n.	协议

III Listening and Speaking

Directions: Scan the QR code on the margin and you will watch a Ted talk entitled "A new wireless sensor network for agriculture communities".

Listening preparation—vocabulary

Please study the following new words that are going to appear in the listening material.

winery	*n.*	a place where wine is made
vineyard	*n.*	a piece of land where grapes are grown in order to produce wine; a business that produces wine
irrigate	*v.*	to supply water to an area of land through pipes or channels so that crops will grow
alignment	*n.*	to change sth slightly so that it is in the correct relationship to sth else
semiconductor	*n.*	a solid substance that conducts electricity in particular conditions, better than insulators but not as well as conductors
specification	*n.*	a detailed description of how sth is, or should be, designed or made

Task 1 Multiple choice

Directions: Please watch the first half of the video and answer the following questions.

1. According to the speaker, the new wireless sensor network may enable wineries to (   ).

   A. know how often to irrigate

   B. improve sustainable agriculture

   C. grow better crop

   D. all above.

2. Which book is written by Jules Verne, the father of science fiction, and mentioned in the talk? (   )

   A. *2001: A Space Odyssey*

   B. *Around the World in Eighty Days*

   C. *Pebble in the Sky*

   D. *Treasure Island*

3. What is true about Arduino according to the speaker? (   )

   A. It was named after a local cafe where the developers often visit.

   B. It was a small computer developed as a project for students.

   C. It has nothing to do with sensors.

   D. One needs to purchase license to use Arduino.

4. The new wireless radio chip—LoRa developed by Semtech differs from WiFi and Bluetooth in that (   ).

A. you can download videos

B. it needs to be recharged every day

C. it communicates over a long range

D. it occupies a high bandwidth

5. How many gateways does LoRa need to provide full wireless coverage in the city of Amsterdam? (   )

A. 15.

B. 3.

C. 10.

D. 200.

Task 2

Directions: Please watch the video and fill in the blanks to complete the note.

A brief history of Internet of the outdoor things

2003	• _____ board available • _____ becomes the standard in this environment and people freely share _____ and _____
2013	• Semtech developed the wireless technology • Sensor may communicate over _____ • With _____ power
2015	• _____ specification • Network provider

Task 3 Spot dictation

Directions: Listen to the talk from 10:05～11:40 and fill in the blanks with the exact words you hear. You might need to fill in more than one word in a blank.

Using the Arduino platform and LoRa as a wireless technology, we have achieved a good coverage of most of Temecula wine country. And that means that you can place a (1) _____ anywhere where there's coverage and just switch it on, and then connect to the network. And the data is (2) _____ on the Internet. We provide about ten miles range, and there's some spotty. So this technology works with (3) _____. So what you can see here is just the coverage where we have very (4) _____. But we cover spots between Winchester and La Cresta. And so this is up to fifty miles. Now this is a community network, and you cannot build a community network without (5) _____ from community members. And I want to specifically mention and call out some of the (6) _____ that really will have been helpful, realizing this network, which can be used by everybody. First of all, I want to mention the range of California water (7) _____, which has been really helpful in supporting this. And although we have (8) _____ wineries use this technology, I want to mention three that actually helped by putting (9) _____ on the roof of their wineries. So that provides up to ten miles of range and free (10) _____ for everybody.

Task 4   Presenting a research paper

As a researcher, you constantly need to attend conferences to get to know the latest research results. You may also need to present your paper and exchange ideas. The Ted Talk you have just watched was about the application of WSN in agriculture. Please search the data base to find a paper on topics related to the application of WSN in other fields. Read the paper and prepare a 5-minute presentation to the class. You should summarize the research questions, research methods and key findings.

Ⅳ Writing

1. Organization of a typical research article

A typical research article often includes title, author information, abstract, key words, body and references. The body part is normally made of introduction, research methods, discussion and result, and conclusion.

2. Title engineering

A piece of academic writing normally contains title, author information, abstract, key words, introduction, body, discussion and references. In this unit, you will learn to write a title.

A title should use the fewest possible words that adequately describe the contents of the paper. It is the most read part in a whole paper. Waste words such as "investigations on" "studies on", or "observations on" should be deleted. Make sure you always start with key words. Effective titles should identify the main issue of the paper, begin with the subject of the paper, are accurate, unambiguous, specific and complete. It should not contain abbreviations, unless they are well-known by the target audience.

Task 1

Directions: Please tick the better title.

(1) A. QR code based chipless RFID system for unique identification
    B. An investigation on QR code based chipless RFID system for unique identification
(2) A. Can UHF RFID be used to manage library materials?
    B. Design of Internet of Things system for library materials management using UHF RFID
(3) A. Studies on the application of IoT to make city smarter
    B. Internet of Things for smart cities

Task 2   Writing your own title

Directions: Here are a range of topics you may choose from to write your own research paper. Please write a title based on those topics.

- Internet of Things and its application in logistics, home, industry and vehicles
- The key enablers of IoT
- Ethical considerations on IoT for the society
- Policy issues about IoT for the government

3. Writing an abstract

As the title is probably the most often read part of your paper, the abstract is the second one. It is a one paragraph summary of the whole paper. In this very short paragraph, you need to state your motivation, problem statement, research methods, results and conclusions.

Motivation means why do you care about the problem and the results. Problem statement is about what problem the paper is trying to solve and what the scope of the work is. Research methods indicate what was done to solve the problem. Results show what the answer to the problem is. Finally, conclusions indicate what implications the answer provides.

There are three types of abstracts: informative, indicative, and the mix of the two. An informative abstract extracts everything relevant from the paper, such as primary research objectives addressed, methods employed in solving the problems, results obtained, and conclusions drawn. Thus it is a highly aggregated substitute for the full paper. While an indicative abstract describes the content and serves as an outline. It cannot serve as a substitute for the paper.

Many researchers prefer to write the abstract after they write the whole paper, so that they are able to write an informative abstract. But you can also write an indicative one before you write the whole paper or any part of it, and use the indicative abstract as an outline or research design guiding your paper writing.

Task 3

Directions: Based on the title you have chosen in Task 2, write an indicative abstract of 100~150 words.

Chapter 1 IoT Technology Fundamentals | 35

KEY

I Reading

Task 1 (tentative answer)

(1) Figure 3-1 shows the hardware structure of WSN sensor node. Power needs to be distributed among sensor, microcontroller and transceiver, while the microcontroller also manages power(as is shown by the two way arrow).

(2) Figure 3-2 shows that the open properties of wireless channels may cause conflicts in time, space or frequency dimension.

(3) Figure 3-3 tells us that sensors are first placed into appropriate places. Second, they wake up and detect data around them. Third, they connect each other into a network. Finally, they may start routing and transmit data.

Task 2 (tentative answer)

(1) The power module provides constant and trustworthy power needed for the system. The sensor joins a WSN node which can get the environmental and equipment status.

(2) … the access network may delay the development of the entire network system.

(3) As the energy is limited in the sensor network environment, not every node is able to transmit data to the sink node due to battery power, processing ability, storage capacity and communication bandwidth.

Task 3

sensor nodes	The hardware of a sensor node generally includes four parts: the power and power management module, a sensor, a microcontroller and a wireless transceiver.
the power module	The power module offers the reliable power needed for the system.
access network technologies	The function of access network technologies is to manage and coordinate the use of channels resources to ensure the interconnection and communication of multiple users on the shared channel.
multi-hop transmission mode	The sensor network nodes are both transmitter and receiver. The first sensor network node, the source node, sends data to a nearby node for data transmission to the gateway. The nearby node forwards the data to one of its nearby nodes that are on the path towards the gateway. The forwarding is

	repeated until the data arrives at the gateway, the destination.
data aggregation	Data aggregation is the process of integrating multiple copies of information into one copy.

Ⅱ Vocabulary

Task

1. from   2. into   3. with   4. as   5. from   6. into   7. by   8. with

Ⅲ Listening and Speaking

Task 1

1. D   2. B   3. B   4. C   5. C

Task 2

2003	• Arduino board available • Open source becomes the standard in this environment and people freely share ideas and programs
2013	• Semtech developed the wireless technology • Sensor may communicate over a long range • With battery power
2015	• LoRa Alliance specification • Network provider

Task 3

(1) sensor   (2) available   (3) line of sight   (4) reliable coverage   (5) collaboration
(6) organizations   (7) district   (8) multiple   (9) gateways   (10) access

Ⅳ Writing

Task 1

(1) A   (2) B   (3) B

Script for Listening (partial, from 00:00～11:45)

Ted X Temecula

Chapter 1 IoT Technology Fundamentals

A new wireless sensor network for agriculture communities

Reinier van der Lee

Like many people, I can appreciate a glass of good well-made wine. It's to Temecula after all. I also like technology. So wine, technology. It makes an interesting combination. Don't you think? This talk is about a new wireless sensor network owned and operated by wineries here in that macula valley. Using technology, we help improve irrigation and improve sustainable agriculture in our whole region. Using the network, we collect sensor data from affordable sensors installed in vineyard. And we use that to provide online guidance to farmers for optimal irrigation, meaning the system will tell you how much and how often to irrigate. That leads to better crops results, and eventually also to better wine, which is my interest here.

But before you go into that, I would like to tell you about the silent computer revolution that happened over the last fifteen years, and the alignment of technologies that happened and made it possible to make the network that we have today in our valley. And for that, I would like to take you on a virtual trip to Europe, to China. We even go to space. And then finally, we'll be coming back to California. And we'll end up beer back home into Temecula. So let's start our industrial history tour.

I grew up in the Netherlands, in an age filled with hippie idealism and a strong belief that we could resolve big world issues with the help of technology. I devoured books by authors like Jules Verne, also considered as the father of science fiction. And I was reading stories about deep sea exploration, traveling to the moon and back here alive, and traveling the world through the air in airships.

What was considered absurdity a century before when these books were written became reality before an early in my lifetime. So from my youth in the Netherlands, let's now go to Italy. We go to a city called Ivrea in the northwestern part of Italy, where in 2003, a small computer was developed as a project for students. But more importantly for this story, it's also the name of a local bar where the developers used to hang out at. And the story goes that they came up with that name under the influence of a couple of beers.

Now this device interacts with its environment using sensors. So it made very well, since it was easy to work with. And it had, you know, support for sensors. So it was an ideal platform for agricultural automation. Soon people all over the world started using Arduino, and they freely shared ideas and programs that other people could use to kick start their own projects. One of the main advantages or the reasons for the success of the project was that it was made available as open source.

And open source means that you can manufacture the Arduino boards and distributed software, free. So it's like permitted to do that. You are permitted to do that. And this triggered the open source maker movements, also triggered by low cost boards coming from China, mainly from the city of Tianjin, which I considered nowadays the open source capital of the world. You can buy and order in a board for less than dollars than ten dollars [SIC], which is before the taxes or whatever Trump is doing right now with.

And if you search for Arduino, we know on Amazon, you also find a wide range of sensors available. Now from Italy, let's go to California, where in 2013, a semiconductor company named Semtech, headquartered in Camarilla, introduced a new wireless radio chip that allowed sensors to communicate over a long range, four times longer than what was available with conventional technologies while on battery power. What made LoRa different than [SIC] a WiFi and the Bluetooth and seller networks is that it trades speed for low power and long range. So you cannot use this wireless technology to download videos. But what you can do with it is monitor a sensor that is five miles away, while that sensor is running all the battery for more than a year.

Now try to do that with your cell phone. I can do it. I usually need to recharge, once a day. What was also added in 2015 was an alliance. And I am with the jack. Uh, if you try to have a network with thousand devices that try to communicate, you have a wicked situation. And so what the LoRa Alliance did, they published a specification that basically, provides a prescription how these devices need to behave to communicate amongst each other in the network.

And now the moment where you all have been waiting for, we go to space the final frontier. This is an artistic impression of the NORSAT-2 satellite. It's a satellite that's orbiting the earth at four hundred miles. So it's low orbit. And it was used for a proof of concept where they sent LoRa signals from space to earth. Why is this important? This goes to show that you can use space technology to make LoRa networks. Or you can place LoRa sensors in remote areas or developing countries.

And there are a number of companies that are actively working in the space, now at this moment. From space, let's go back to Europe. And we land here in Amsterdam. We were in 2015: a kick starter project named The Things Network, had a project where they had sensors installed in boats that were moored in the canals of Amsterdam. And these sensors were supposed to detect leaks during the winter period. Now what was remarkable about this project was that they only needed ten gateways or access points to provide full LoRa wireless coverage of the city of Amsterdam.

The Things Network provides free to use community networks where the owners or the users own the network. And this concept has exploded since 2015, so it's just three years. And there are now nearly 5,000 community networks worldwide with more than 50,000 users. With all this in place, you know, I have the background and I just want to reiterated the most important milestones for the remainder of my story. First, in 2003, there was this open source, Arduino boards available. Open source became the standards in this environment and people freely shared ideas and programs and projects for other people to build upon. In 2013, we got this wireless technology that allowed sensors to communicate over long distance following battery power.

And finally, there was a specification that tamed the conversation between sensors and allowed network providers to provide sensor networks that, you know, people can subscribe to it. So with these three key ingredients in place, I can now take you back to Temecula, and the network that we have realized here to support our wineries and grape growers. This is a picture of the home that I bought in 2010 together with my wife. I must admit it took some persuasion, but yeah, we got it. And then we were confronted with a multi-year drought soon afterwards, so that brought me on the fast lane educational paths on how to irrigate the vineyard and how to improve that.

Now Temecula has seen a tremendous growth in the number of wineries, and they're still, you know, new wineries added as we speak. They also have seen an increasing in quality of grapes over the last 50 years since its inception to make your wine valley. So the saying goes, good quality wine starts in the vineyard. And good irrigation is important for quality fruit. Using the Arduino platform and LoRa as a wireless technology, we have achieved a good coverage of most of Temecula wine country.

And that means that you can place a sensor anywhere where there's coverage and just switch it on, and then connect to the network. And the data is available on the Internet. We provide about ten miles range, and there's some spotty. So this technology works with line of sight. So what you can see here is just the coverage where we have very reliable coverage. But we cover spots between Winchester and La Cresta. And so this is up to fifty miles. Now this is a community network, and you cannot build a community network without collaboration from community members.

And I want to specifically mention and call out some of the organizations that really will have been helpful, realizing this network, which can be used by everybody. First of all, I want to mention the range of California water district, which has been really helpful in supporting this. And although we have multiple wineries use this technology, I want to mention three that actually helped by putting gateways on the roofs of their wineries. So that provides up to ten miles of range and free access for everybody. Those three are… and I'll do it in alphabetical sequence. Otherwise I get in trouble.

Unit 4　Machine-to-Machine

1　M2M (Machine-to-Machine) technology was first adopted in manufacturing and industrial settings, and later found applications in healthcare, business, insurance and more. It is also the foundation for the Internet of things (IoT).

History of machine-to-machine technology

2　The roots of M2M are planted firmly in the manufacturing industry, where other technologies, such as **SCADA** and remote monitoring, helped remotely manage and control data from equipment.

3　While the origins of the acronym are unverified, the first use of machine-to-machine communication is often credited to Theodore Paraskevakos, who invented and patented technology related to the transmission of data over telephone lines, the basis for modern-day caller ID.

4　Nokia was one of the first companies to use the acronym in the late 1990s. In 2002, it partnered with Opto 22 to offer M2M wireless communication services to its customers.

5　In 2003, M2M Magazine launched. The publication has since defined the six pillars of M2M as remote monitoring, RFID, **sensor** networking, smart services, **telematics** and **telemetry**.

How M2M works

6　The main purpose of machine-to-machine technology is to tap into sensor data and transmit it to a network. Unlike SCADA or other remote monitoring tools, M2M systems often use public networks and access methods—for example, **cellular** or **Ethernet**—to make it more cost-effective.

7　The main components of an M2M system include sensors, RFID, a Wi-Fi or cellular communications link, and autonomic computing software programmed to help a network device interpret data and make decisions. These M2M applications translate the data, which can trigger preprogrammed, automated actions.

---

setting /ˈsetɪŋ/ *n.*
the state of the environment in which a situation exists

acronym /ˈækrənɪm/ *n.*
a word formed from the initial letters of the several words in the name

unverified /ʌnˈverɪfaɪd/ *adj.*
lacking proof or substantiation

be credited to owe to

patent /ˈpætnt/ *v.*
to obtain a document granting an inventor sole rights to an invention

transmit /trænzˈmɪt/ *v.*
the act of sending a message; causing a message to be transmitted

transmission /trænzˈmɪʃən/ *n.*

pillar /ˈpɪlə/ *n.*
fundamental principle or practice

tap into to dig/exploit

component /kəmˈpəʊnənt/ *n.*
an abstract part of something

autonomic /ˌɔːtəˈnɒmɪk/ *adj.*
occurring involuntarily or spontaneously

automated /ˈɔːtəmeɪtɪd/ *adj.*
to use machines and computers instead of people to do a job or task

8   One of the most well-known types of machine-to-machine communication is telemetry, which has been used since the early part of the last century to transmit <u>operational</u> data. Pioneers in telemetrics first used telephone lines, and later, radio waves, to transmit performance <u>measurements</u> gathered from monitoring instruments in remote locations.

9   The Internet and improved standards for wireless technology have expanded the role of telemetry from pure science, engineering and manufacturing to everyday use in products such as heating units, electric meters and Internet-connected devices, such as <u>appliances</u>.

10   Beyond being able to remotely monitor equipment and systems, the top benefits of M2M include:
- reduced costs by minimizing equipment maintenance and <u>downtime</u>;
- boosted <u>revenue</u> by revealing new business opportunities for servicing products in the field;
- improved customer service by <u>proactively</u> monitoring and servicing equipment before it fails or only when it is needed.

M2M applications

11   Machine-to-machine communication is often used for remote monitoring. In product <u>restocking</u>, for example, a <u>vending</u> machine can message the distributor's network, or machine, when a particular item is <u>running low</u> to send a refill. An enabler of <u>asset</u> tracking and monitoring, M2M is <u>vital</u> in warehouse management and supply chain management.

12   <u>Utilities</u> companies often rely on M2M devices and applications to not only <u>harvest</u> energy, such as oil and gas, but also to bill customers—through the use of smart meters—and to detect worksite factors, such as pressure, temperature, equipment status and more.

13   In telemedicine, M2M devices can enable the real-time monitoring of patients' vital statistics, <u>dispensing</u> medicine when required, or tracking healthcare assets. M2M is also an important aspect of remote control, <u>robotics</u>, traffic control, security, logistics and <u>fleet</u> management, and automotive.

M2M vs. IoT

14   While many use the terms interchangeably, M2M and IoT

---

operational /ˌɒpəˈreɪʃənəl/ *adj.*
pertaining to a process or series of actions for achieving a result
measurement /ˈmɛʒəmənt/ *n.*
a result, usually expressed in numbers, that you obtain by measuring something
appliance /əˈplaɪəns/ *n.*
a device or machine designed to do a particular thing in the home, such as heating, cleaning or cooking (often electrical)
downtime /ˈdaʊnˌtaɪm/ *n.*
the time during which machinery or equipment is not operating
revenue /ˈrɛvɪnjuː/ *n.*
money that a company, organization, or government receives from people
proactive /prəʊˈæktɪv/ *adj.*
controlling a situation by making things happen rather than waiting for things to happen and then react to them
proactively *adv.*
restock /riːˈstɒk/ *v.*
to fill with goods to replace what has been used or sold
vend /vɛnd/ *v.*
to sell or be sold
run low to be spent or finished
asset /ˈæsɛt/ *n.*
a person or thing that is valuable or useful to sb/sth
vital /ˈvaɪtəl/ *adj.*
urgently needed; absolutely necessary
utility /juːˈtɪlɪtɪ/ *n.*
important services such as water, electricity, or gas that is provided for everyone, and that everyone pays for
harvest /ˈhɑːvɪst/ *v.*
gather, as of natural products
dispense /dɪˈspɛns/ *v.*
to prepare medicine and give it to people
robotics /rəʊˈbɒtɪks/ *n.*
the science of designing and building robots
fleet /fliːt/ *n.*
a group of vehicles, especially when they all belong to a particular organization or business, or when they are all going somewhere together

Vocabulary	
disparate /ˈdɪspərɪt/ *adj.*	clearly different from each other in quality or type
ecosystem /ˈiːkəʊˌsɪstəm/ *n.*	all the plants and animals that live in a particular area together with the complex relationship that exists between them and their environment
enterprise /ˈɛntəˌpraɪz/ *n.*	a company or business
incorporate /ɪnˈkɔːpəreɪt/ *adj.*	formed or united into a whole
predict /prɪˈdɪkt/ *v.*	make a prediction about; tell in advance
personalize /ˈpɜːsənəˌlaɪz/ *v.*	make personal or more personal
personalized *adj.*	
unauthorized /ʌnˈɔːθəˌraɪzd/ *adj.*	not endowed with authority; without official authorization
access /ˈækses/ *n.*	the right to obtain or make use of or take advantage of something (as services or membership)
intrusion /ɪnˈtruːʒən/ *n.*	entrance by force or without permission or welcome
fraud /frɔːd/ *n.*	intentional deception resulting in injury to another person
encryption /ɛnˈkrɪpʃən/ *n.*	the activity of converting from plain text into code
segment /ˈsɛgmənt/ *v.*	divide or split up
confidentiality /ˌkɒnfɪˌdɛnʃɪˈælɪti/ *n.*	discretion in keeping secret information
combat /ˈkɒmbæt/ *v.*	fight against sth or try to stop sth from happening

are not the same. IoT needs M2M, but M2M does not need IoT.

15 Both terms relate to the communication of connected devices, but M2M systems are often isolated, stand-alone networked equipment. IoT systems take M2M to the next level, bringing together <u>disparate</u> systems into one large, connected <u>ecosystem</u>.

16 M2M systems use point-to-point communications between machines, sensors and hardware over cellular or wired networks, while IoT systems rely on IP-based networks to send data collected from IoT-connected devices to gateways, the cloud or middleware platforms.

17 Data collected from M2M devices is used by service management applications, whereas IoT data is often integrated with <u>enterprise</u> systems to improve business performance across multiple groups. Another way to look at it is that M2M affects how businesses operate, while IoT does this and affects **end users**.

18 For example, in the product restocking example above, M2M involves the vending machine communicating to the distributor's machines that a refill is needed. <u>Incorporate</u> IoT and an additional layer of analytics is performed; the vending machine can <u>predict</u> when particular products will need refilling based on purchase behaviors, offering users a more <u>personalized</u> experience.

M2M security

19 Machine-to-machine systems face a number of security issues, from <u>unauthorized</u> <u>access</u> to wireless <u>intrusion</u> to device hacking. Physical security, privacy, <u>fraud</u> and the exposure of mission-critical applications must also be considered.

20 Typical M2M security measures include making devices and machines **tamper-resistant**, embedding security into the machines, ensuring communication security through <u>encryption</u> and securing back-end servers, among others. <u>Segmenting</u> M2M devices onto their own network and managing device identity, data <u>confidentiality</u> and device availability can also help <u>combat</u> M2M security risks.

M2M standards

21 Machine-to-machine technology does not have a standardized

device platform, and many M2M systems are built to be task-or device-specific. Several key M2M standards, many of which are also used in IoT settings, have emerged over the years, including:
- **OMA** DM (Open Mobile Alliance Device Management), a device management protocol
- OMA Lightweight M2M, a device management protocol
- **MQTT**, a messaging protocol
- TR-069 (Technical Report 069), an application layer protocol
- HyperCat, a data discovery protocol
- OneM2M, a communications protocol
- Google Thread, a **wireless mesh** protocol
- AllJoyn, an open source **software framework**

Word count: 920

Source: Adapted from Shea S. Machine-to-Machine (M2M) [EB/OL]. [2019-02-01]. https://internetofthingsagenda.techtarget.com/definition/machine-to-machine-M2M.

> alliance /əˈlaɪəns/ n.
> a group of countries or political parties that are formally united and working together because they have similar aims
>
> protocol /ˈprəʊtəˌkɒl/ n.
> a set of rules for exchanging information between computers
>
> mesh /meʃ/ n.
> the topology of a network whose components are all connected directly to every other component
>
> framework /ˈfreɪmˌwɜːk/ n.
> a particular set of rules, ideas, or beliefs which you use in order to deal with problems or to decide what to do

Ⅰ Reading

Reading for a purpose

When reading magazines, we read for fun; while reading novels, we read for plots. If they are news reports, we pay attention to what happened; if academic papers, we care much about object, methods and conclusion. When it comes to textbooks, we pay extra attention to what we called "Knowledge Node"—the key points or the important segments of this piece of knowledge.

We always read for some kinds of purposes. Often it's the information we are interested in or we need to improve our knowledge hierarchy; or some other times, to finish certain exercises. Whatever your purpose may be, remembering some basic information like Who, When, Where, What and How is a premise to further understanding of your article.

1. Recalling information

Basic information about Who, When, Where, What and How are the bones of the article. Recalling them can help forming a structure of it.

Task 1

Directions: Please recall the basic information about the history of M2M and fill in the following blanks.

- Inventor: _____
- The origins of the acronym M2M: _____
- One of the first companies to use the acronym M2M: _____
- First time offering services to customers in _____
- M2M Magazine was launched in _____.
- The six pillars of M2M are _____.

2. Reading for specific details

Task 2

Directions: Please answer the following questions.

1. What is the main purpose of machine-to-machine technology?

2. In what way have the Internet and improved standards for wireless technology pushed the development of telemetry?

3. What are some of the benefits which M2M brings us?

4. What's the difference between M2M and IoT in terms of the usage of collected data?

5. What are some typical M2M security measures?

Ⅱ Vocabulary

Key words and expressions

setting	acronym	unverified	be credited to
patent	transmission	pillar	tap into
component	autonomic	automated	operational
measurement	appliance	downtime	revenue
proactive	restock	vend	run low
asset	vital	utility	harvest
dispense	robotic	fleet	disparate
ecosystem	enterprise	incorporate	predict
personalize	unauthorized	access	intrusion
fraud	encryption	segment	confidentiality
combat	alliance	protocol	mesh
framework			

Terminology

SCADA (supervisory control and data acquisition)	n.	监控与数据采集
sensor	n.	传感器
telematics	n.	远程信息处理
telemetry	n.	遥测技术；遥感勘测
cellular	n.	蜂窝网络
Ethernet	n.	以太网
end user	n.	终端用户
tamper-resistant	adj.	防篡改的
OMA (Open Mobile Alliance)	n.	开放移动联盟
MQTT (message queuing telemetry transport)	n.	消息队列遥测传输
wireless mesh	n.	无线网状网
software framework	n.	软件框架

Task 1 Word bank

Directions: Please fill in the blanks with the words listed below. You may need to change the form when necessary.

unverified	vital	predict	personalize
component	access	combat	segment
operational	incorporate	dispense	intrusion

1. I believe there will be one day in near future when people can find that wild animals

play a _____ role in keeping the balance of the nature.

2. These resources are not visible or _____ from any other virtual portal and therefore cannot be shared among virtual portals.

3. Published but _____ reports indicate that the corporation employs at least 25,000 people, many of whom live in the urban area of about 30,000 that exists outside Huaxi's cramped legal boundaries.

4. This is particularly concerning since personal data is increasingly used to make _____ about ourselves and our potential future behavior.

5. What he had to do now was to analyze the new material and _____ it into his profile.

6. A particular _____ of music reminds you of a part of your life, your experiences, your childhood memories.

7. You may only strike with one weapon per _____, but you can strike with it multiple times if you find a way to ready it.

8. Health officials hope to begin _____ anti-retroviral (抗反转录病毒) drugs on a wide scale at the beginning of next year.

9. Although the original body parts and other _____ were virtually all intact, the car had lost its engine and gearbox.

10. It is more likely that you get more _____ services in small companies.

Task 2 Word formation

In linguistics, word formation is the creation of a new word on the basis of other words or morphemes. There are many ways to form a new word. The following are some examples.

• Compounding: home+work=homework, birth+control=birthcontrol
• Derivation: happy—happily, regular—regularly, lead—mislead
• Conversion: hand(*n.* 手, *v.* 上交), land(*n.* 土地, *v.* 着陆), cook(*n.* 厨师, *v.* 烹饪)
• Blending: breakfast+lunch=brunch, Chinese+English=Chinglish
• Clipping: refrigerator—fridge, hamburger—burger
• Acronym: People's Republic of China—PRC, be right back—BRB, laughing out loud—LOL

Directions: *Please fill in the following chart with relative words; the grey ones should be ignored.*

Words in the text	*n.*	*v.*	*adj.*	*adv.*
measure				
unauthorized				
appliance				
operational				
proactively				

Ⅲ Listening and Speaking

Directions: Scan the QR code on the margin and you will watch a video of an advertisement of Vodafone's M2M service.

Listening preparation—vocabulary

Video Clip

Please study the following new words which are going to appear in the listening material.

recognition	*n.*	the act of recognizing someone or identifying something when you see it
intervention	*n.*	the act of becoming involved in something and trying to change it
go off		be activated
commute	*v.*	to travel regularly by bus, train, car, etc. between your place of work and your home
dynamic	*adj.*	the way in which people or things behave and react to each other in a particular situation
utility	*n.*	a service provided for the public, for example an electricity, water or gas supply
asset	*n.*	all the things that a company or a person owns
consult	*v.*	to go to sb for information or advice
lease	*v.*	to use or let sb use sth, especially property or equipment, in exchange for rent or a regular payment
proactive	*adj.*	controlling a situation by making things happen rather than waiting for things to happen and then reacting to them
agility	*n.*	the gracefulness of a person or animal that is quick and nimble
engagement	*n.*	being involved with sb/sth in an attempt to understand them/it

Task 1 Multiple choice

Directions: Please watch the video and answer the following questions by choosing the right answers from the four choices marked A, B, C and D.

1. Which of the following home services is NOT provided by M2M? (   )

A. To change the alarming time of clock automatically.

B. To help cancel the train for you if you are asleep.

C. To turn on the heat when it's cold outside.

D. To help monitor your health remotely.

2. Suppose you're a senior student in college, while driving on an M2M technology powered road, which of the following advertisement you would most probably see alongside the road? (   )

A. An advertisement of a suburban villa.

B. An advertisement of a housekeeping service sites.

C. An advertisement of a job site.

D. An advertisement of a limousine.

3. Which of the following is NOT a business service provided by M2M?(   )

A. It can monitor and analyze your business energy usage.

B. It may give you the information you need to save on utility costs.

C. It can help monitor your assets remotely to let you know the status of them in real time.

D. It can help you fix your machine when they break down.

4. According to this advertisement, who could Vodafone bring benefit to?(   )

A. Business managers.

B. Housewives.

C. The more engaged customers.

D. People in all professions.

Task 2   Spot dictation

Directions: Please listen to the part (01:44~03:00) of the video, and fill in the blanks below. You may listen to the video several times to get the correct words.

The benefits(1) _____ when Mark gets to work. M2M technology can monitor and analyze Mark's business energy usage, giving Mark the information he needs to save on (2) _____ costs. When it comes to Mark's business assets, from fleets of trucks to vending machines, remote monitoring lets him know the status of them in real time. And if a problem (3) _____, the machine can be fixed before they (4) _____ saving time and money. Just imagine what services you could offer your customers when you know what your machines are doing. New business modes are now a real possibility: consulting, leasing, (5) _____ servicing and more. Also, with assets (6) _____, Mark has the power to know the location of his assets, such as a fleet of delivery trucks or the products being transported in them, with all their status information (7) _____ at his (8) _____. Best of all, this can be achieved on a global scale thanks to Vodafone's worldwide networks. So whatever profession you're in, M2M services from Vodafone can allow your business to (9) _____, meaning you could do much more with much less: less cost, less energy and less time; more growth, more visibility and more control, with happier and more (10) _____ customers.

Task 3   Presenting an academic problem with the help of PowerPoint
           or Keynote

Directions: Please select a term, a thing or a problem related to IoT, and make several slides to introduce it to your classmates. You may imitate the mode and procedure in the video and you should make it clear and easy to understand.

Ⅳ Writing

1. Writing a research paper: body part—introduction

Every research paper needs context so that readers can understand why you have created it. This is exactly what you can do in the introduction part. Of course, this can mean that your introduction is the hardest part of paper to write first. So, it is essential that you take your time and make sure that you get it right.

In writing the introduction, you should begin with background information on the topic; you should narrow it down to specific points, make it clear to the audience and stress the connection with the problem; but you should not include research details in the introduction.

Then you should conduct literature review to state valuable opinion. That means you should collect the works of people who work in the area of your interest, retrieve some quotations to insert as in-text citations, and develop a corresponding reference to each source. Your introduction to a research paper reflects what the literature was about and how it helped to explore the selected topic.

And after that, what else should be included in your introduction? You should also:
- introduce your topic;
- tell your readers about the research you plan to carry out;
- state your rationale (meaning why you wrote this paper);
- explain why your research is important;
- state your hypothesis (if there is).

Following these steps, you may get yourself a good introduction.

2. Writing a research paper: body part—literature review

In writing a literature review of a research paper, one should have a preface, a body part, a summary and references.

The preface should use concise words to explain the purpose, necessity, definition of relevant concepts, the scope of the review, the status quo and dynamics of the relevant issues, and the current focus on the main issues. The preface is generally 200~300 words, and should not exceed 500 words.

The body part is the focus of the review. There is no fixed format in the writing. As long as the comprehensive content can be expressed well, the author can creatively adopt many forms. The body part should include the argument and the argumentation, by asking questions, analyzing and solving problems, comparing different scholars' views on the same issue and their theoretical basis, further clarifying the ins and outs of the problem and the author's own opinions. Of course, the author can also present different views of the literature from the historical background, current status, and development direction of the problem. The body part can be divided into smaller parts under sub-headings according to the content.

The author should then conduct a comprehensive summary to evaluate various viewpoints, put forward his own views, point out the existing problems and the direction and prospects for future development. A simple review of the content can also be done without a summary.

References are an important part of the review, too.

(https://edubirdie.com/blog/research-paper-introduction

https://writemypaper4me.org/blog/research-paper-introduction)

Task 1 Write an introduction

Direction: Please write an academic introduction of what you have introduced in Task 3, Part III for 100~200 words.

Task 2 Write a mini-literature review

Direction: Please refer to the Internet, find existing articles and studies about what you have written about in Task 1, and write a mini-literature review for 100~200 words.

KEY

I Reading

Task 1

- Inventor: Theodore Paraskevakos
- The origins of the acronym M2M: unverified
- One of the first companies to use the acronym M2M: Nokia
- First time offering services to customers in 2002
- M2M Magazine was launched in 2003

- The six pillars of M2M are <u>remote monitoring, RFID, sensor networking, smart services, telematics and telemetry</u>

Task 2

1. What is the main purpose of machine-to-machine technology?
<u>To tap into sensor data and transmit it to a network.</u>

2. In what way have the Internet and improved standards for wireless technology pushed the development of telemetry?
<u>They have expanded the role of telemetry from pure science, engineering and manufacturing to everyday use in products such as heating units, electric meters and Internet-connected devices, such as appliances.</u>

3. What are some of the benefits which M2M brings us?
<u>a. reducing costs by minimizing equipment maintenance and downtime; b. boosting revenue by revealing new business opportunities for servicing products in the field; and c. improving customer service by proactively monitoring and servicing equipment before it fails or only when it is needed.</u>

4. What's the difference between M2M and IoT in terms of the usage of collected data?
<u>Data collected from M2M devices is used by service management applications, whereas IoT data is often integrated with enterprise systems to improve business performance across multiple groups.</u>

5. What are some typical M2M security measures?
<u>Typical M2M security measures include making devices and machines tamper-resistant, embedding security into the machines, ensuring communication security through encryption and securing back-end servers, among others. Segmenting M2M devices onto their own network and managing device identity, data confidentiality and device availability can also help combat M2M security risks.</u>

II Vocabulary

Task 1

1. vital 2. accessible 3. unverified 4. predictions 5. incorporate
6. segment 7. combat 8. dispensing 9. components 10. personalized

Task 2

Words in the text	n.	v.	adj.	adv.
measure	measure	measure	measurable	measurably
unauthorized	authorization	authorize	authorized	
appliance	application	apply	applied/applicable	
operational	operation	operate	operable	operationally
measurement	measure	measure	measurable	measurably
proactively	activity	activate	active	actively

Ⅲ Listening and Speaking

Task 1
1. B 2. C 3. D 4. D

Task 2
(1) continue (2) utility (3) arises (4) break down (5) proactive
(6) tracking (7) instantly (8) fingertips (9) innovate (10) engaged

Script for Listening

In the past generation, the way we communicate has transformed beyond all recognition. Today the world is more connected than ever. Now technology is connecting not just people, but machines and devices together on a global scale, forever changing the way we do business. And Vodafone M2M is driving this revolution.

But what is M2M? You might know it as the Internet of Things. But M2M stands for Machine to Machine. It's a technology that allows physical objects or machines to connect to the Internet. This means they can communicate and share information over the Vodafone network without the need for human intervention. But how can M2M benefit our lives and businesses?

Meet Mark. Mark is fast asleep in his smart home that's powered by M2M technology. Mark is just like you and me. He loves to sleep. But today Mark's trains are canceled, meaning he will need to drive to work instead. Luckily for Mark, his connected alarm clock automatically goes off ten minutes early, ensuring he won't be late. And best of all, Mark didn't even have to do a thing. And with smart sensors installed, Mark's home automatically adjusted to its environment. So if it's a cold day, Mark doesn't have to worry about turning on the heating. His home does it for him. These are just some of the multitude of smart home services M2M offers along with everything from remotely monitored security alarms to remote health monitoring.

And that's not all. As he commutes to the office, Mark finds that M2M technology is everywhere, from real time adapted traffic control systems to smart ways management, and dynamic advertising content systems which target individual customers.

The benefits continue when Mark gets to work. M2M technology can monitor and analyze Mark's business energy usage, giving Mark the information he needs to save on utility costs. When it comes to Mark's business assets, from fleets of trucks to vending machines, remote monitoring lets him know the status of them in real time. And if a problem arises, the machine can be fixed before they break down, saving time and money. Just imagine what services you could offer your customers when you know what your machines are doing. New business modes are now a real possibility: consulting, leasing, proactive servicing and more. Also, with assets tracking, Mark has the power to know the location of his assets, such as a fleet of delivery trucks or the products being transported in them, with all their status information instantly at his fingertips. Best of all, this can be achieved on a global scale thanks to Vodafone's worldwide networks. So whatever profession you're in, M2M services from Vodafone can allow your business to innovate, meaning you could do much more with much less: less cost, less energy and less time; more growth, more visibility and more control, with happier and more engaged customers.

So let Vodafone show you how M2M solutions can help businesses deliver better operational agility and better customer engagement. M2M Services from Vodafone—Transforming lives and businesses.

Unit 5 Setting the Standard

Introduction

1 Although the Internet of Things (IoT) is seen as a vision of what is to come, rather than a technology in and of itself, it reflects trends in both technological innovation and business strategy. It refers to the convergence of previously disparate **telecommunication** capabilities. For the IoT to become a reality, the development of many different types of technology will have to be coordinated, ranging from item labeling and process control to wireless technology and network interconnection.

2 These requirements are illustrated in Figure 5.1. **Product identification** refers to the mechanisms by which individual items can be identified and tracked, via for instance traditional bar codes or RFID tags. Sensor network and home automation technologies that have developed from industrial process control systems make it possible to monitor the ambient environment. Wireless technology is of course a pre-requisite (enabling any physical object to become a part of this ubiquitous network) as is network interconnection via the Internet (enabling global access and reach). Although wireless access technology will become prevalent, there will continue to be a role for wired systems such as **power line communication (PLC)** within the home.

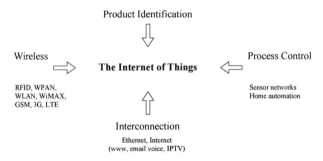

Figure 5.1 The convergence of product identification, process control, wireless and interconnection technology applications

3 The IoT will consist of objects with tags and networked readers, writers, sensors and actuators. The telecommunication systems of today that primarily support interpersonal and person-

to-machine interaction will be enhanced with an increasing array of machine-to-machine communications.

Standardizing the IoT

4   Given the ongoing and emerging convergence of technology areas, a diverse number of organizations from previously separate industry segments are involved in the specification of systems and their standardization. Not surprisingly, this has led to some overlap in activities because these organizations are each working in their own specific area of expertise. The result has been a bewildering array of standards. A 2008 European study on RFID alone noted that more than 250 standards describing RFID-related solutions had been established by around 30 different organizations. In this context, international standardization organizations can play an important role in harmonizing specifications and creating interoperable global standards for the IoT.

5   The most important organizations setting standards for the IoT are:

- EPCglobal, the Ubiquitous ID Center and ISO/IEC (International Organization for Standardization/International Electrotechnical Commission) in the area of defining identifier formats and short-range radio technology;
- The **IEEE** 802 standards committee on local and personal area networks;
- the Internet Engineering Task Force (**IETF**) for the suite of protocols that provide end-to-end connectivity over the Internet.

6   The **ITU-T** (International Telecommunication Union—Telecommunication Standardization Sector) is also playing a role in harmonizing standards and producing system-level description of **ubiquitous networks**.

Why standardize?

7   Standards can be used to increase product quality (i. e. meeting performance and safety requirements) and to ensure the interoperability of various components in a system. It is this latter aspect that is of interest in the present context. Standards are particularly valuable in cases where interfaces between

---

interaction /ˌɪntərˈækʃən/ *n.*
the process of information or instruction exchanging
array /əˈreɪ/ *n.*
a large number or wide range of thing
diverse /daɪˈvɜːs/ *adj.*
very different from each other and of various kinds
specification /ˌspɛsɪfɪˈkeɪʃən/ *n.*
a restriction that is insisted upon as a condition for an agreement
overlap /ˈəʊvəlæp/ *n.*
a representation of common ground between theories or phenomena
expertise /ˌɛkspɜːˈtiːz/ *n.*
special skill or knowledge that is acquired by training, study, or practice
bewilder /bɪˈwɪldə/ *v.*
cause to be confused emotionally
bewildering *adj.*
interoperable /ˌɪntəˈɒpərəbl/ *adj.*
able to exchange and use information
identifier /aɪˈdɛntɪˌfaɪə/ *n.*
symbol that establishes the identity of the one bearing it
format /ˈfɔːmæt/ *n.*
the organization of information according to preset specifications

harmonize /ˈhɑːməˌnaɪz/ *v.*
bring into consonance or accord

components are produced by different companies (whether or not these are physically separate pieces of equipment) or where items of equipment are owned by different organizations. Ideally, standardization should provide <u>mutual</u> benefits for equipment vendors, service providers and their customers, by stimulating the overall growth of a particular market.

8   In general, significant benefits are to be gained by standardization of:
- the information to be transferred, such as the format of the identifier and the application data;
- the characteristics of the interfaces;
- the protocols for data transfer over the various interfaces;
- other functions, such as **routing** and security.

9   In addition to the standardization of interface specifications and protocols, companies may also be required to follow specific regulations. Examples include those concerned with radio frequency usage (e.g. to ensure the interoperability of equipment) or those concerned with the protection of consumers using the technology (e.g. data protection <u>legislation</u> and guidelines).

What needs to be standardized?

10   The IoT can be viewed as a <u>subset</u> of a future Internet in which communication capabilities will become ubiquitous. However, it is widely acknowledged that the IoT suffers from a <u>fragmentation</u> of standards. For examples, EPCglobal, ISO and Japan's Ubiquitous ID organization have defined formats for <u>tag</u> data. At the same time, other organizations have been active in defining local and wide-area network connectivity standards. It is therefore necessary to consider the technology and standards produced in the four areas (see Figure 5.1) that are converging and how these technologies can be integrated in a complete system with end-to-end connectivity. For instance, the standardization of sensor networks is relevant to the broader picture of standardization activities in this area. **Home networking** also provides an example of how RFID, sensor networks, wireless and fixed (e.g. PLC) communication links and the more familiar applications of the Internet might be integrated. Some of the standards relating to ubiquitous networking in next generation networks are relevant in this context.

---

mutual /ˈmjuːtʃuəl/ *adj.*
common to or shared by two or more parties

legislation /ˌledʒɪsˈleɪʃən/ *n.*
the act of making or enacting laws

subset /ˈsʌbˌset/ *n.*
a smaller number of things that belong together within that group

fragmentation /ˌfræɡmenˈteɪʃən/ *n.*
separating something into fine particles

tag /tæɡ/ *n.*
a label attached to the target

11 Figure 5.2 provides a good framework in which to consider the various elements of the IoT. It shows the identifiers, interfaces and some of the wide-area network functions involved in connecting "things" to the Internet. In illustrating how the various technologies can be integrated to create an IoT, it reflects the areas of convergence illustrated in Figure 5.1. It can be used as an effective model for analyzing standardization activities in each area.

12 More specifically, in a typical system an object is assigned a tag with an identifier. In some cases, additional application data can be associated with the object. These application data could, for instance, be provided by a sensor collocated with the tag. The identifier and application data are read over a short-range radio frequency interface, such as RFID, or by a scanner. **ID terminals**, such as readers and sensors, will use low-power wireless networks—networks that can be connected to the global Internet.

> **be associated with** be linked with, be connected with
> **collocate** /ˈkɒləˌkeɪt/ v.
> place side by side
> **terminal** /ˈtɜːmɪnəl/ n.
> a piece of equipment consisting of a keyboard and a screen that is used for putting information into a computer or getting information from it

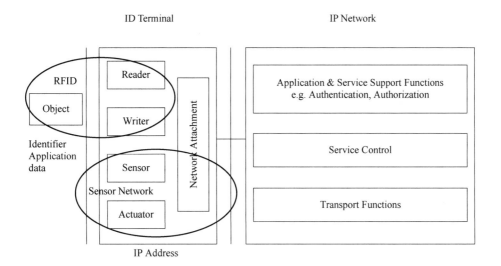

Figure 5.2 Reference model for the IoT, adapted from reference architecture for tag-based applications in ITU-T recommendation Y.2213

13 In summary, therefore, the key areas requiring standardization are as follows: the identification of things, the methods by which information is transferred between things and the devices (ID

terminals) that detect or control them, the networking of ID terminals, and finally, the method by which ID terminals are connected to the global Internet.

Word count: 1,044

Source: Adapted from Mainwarning K, Srivastava L. The Internet of Things—Setting the standards[M]//Chaouchi H. The Internet of Things—Connecting objects to the Web. London: ISTE, 2010: 191-196.

Ⅰ Reading

Searching for academic papers

Before starting with your research, the preliminary job is always reading. Through reading, you may know who are the pioneers in this field, what has done and what is yet to be done; what methods you could apply and what hypothesis you could propose.

But what to read? Searching for relevant, high-quality academic papers is one of the crucial skills in academic reading and writing. In order to be an efficient searcher, you have to pay attention to the following aspects.

- Search engine. Choose a professional academic search engine which has a powerful source and great technology backing.
- Key words. Key words are the KEY. They decide the relevance of the papers you got. Choose multiple key words, and try the combination of them when you search.
- Abstract. Abstract can help you quickly understand what the paper is about. Going over abstracts before serious reading may save you a lot of time.
- Other than the above mentioned, citation ranking, publisher and the year of publication are also aspects you may want to consider.

1. Searching for important information

Task 1

Directions: Please scan the text, locate the following sentences and fill in the blanks.

1. For the IoT to become a _____, the development of many different types of technology will have to be _____, ranging from _____ _____ and _____ _____ to _____ _____ _____ and _____ _____.

2. In this context, international standardization organizations can play an important role in _____ _____ and creating _____ _____ _____ for the IoT.

3. In summary, therefore, the key areas requiring standardization are as follows: the _____ of things, the methods by which information is _____ between things and the

_____ (ID terminals) that _____ or _____ them, the _____ of ID terminals, and finally, the method by which ID terminals are _____ to the global Internet.

2. Reading and understanding the text

Task 2   Multiple choice

Directions: Please answer the following questions by choosing the right answers from the four choices marked A, B, C and D.

1. Why is there a bewildering array of standards? (   )

A. Different organizations were involved in specific systems and their activities overlap.

B. More than 250 standards had been established by around 30 different organizations.

C. The most important organizations setting standards for the IoT are over-doing their works.

D. The ITU-T is not playing an effective role in harmonizing standards and producing system-level description of ubiquitous networks.

2. IoT needs to be standardized because of the following reasons EXCEPT(   )

A. Standards can be used to increase meeting performance and safety requirements.

B. Standards can help ensure the interoperability of various components in different systems.

C. Standards are especially valuable when interfaces are produced by different companies.

D. Standards could help increase market growth and therefore provide mutual benefits for all.

3. Which of the following is one of the most important organizations setting standards for the IoT? (   )

A. the RFID   B. the PLC
C. the ITU-T   D. the IETF

4. What is illustrated in Figure 5.2? (   )

A. It shows how various technologies can be integrated to create the IoT.

B. It shows how the IoT standardization process goes.

C. It shows how the organizations set standards for the IoT.

D. It shows how an object is assigned with a tag which collocated with its application.

5. What is NOT one of the key areas requiring standardization? (   )

A. The identification of things

B. Information transformation between things and their controlling devices

C. Information transformation between detecting and controlling devices

D. The networking of ID terminals and its connection to the global Internet

Ⅱ Vocabulary

Key words and expressions

innovation	convergence	coordinate	mechanism
ambient	requisite	prevalent	interpersonal
interaction	array	diverse	specification
overlap	expertise	bewilder	interoperable
identifier	format	harmonize	mutual
legislation	subset	fragmentation	tag
be associated with	collocate	terminal	

Terminology

telecommunication	n.	电信
product identification	n.	产品识别
PLC (power line communication)	n.	电力线通信
IEEE (Institute of Electrical and Electronics Engineers)	n.	电气和电子工程师协会
IETF (Internet Engineering Task Force)	n.	国际互联网工程任务组
ITU-T (International Telecommunication Union—Telecommunication Standardization Sector)	n.	国际电信联盟电信标准分局
ubiquitous networks	n.	泛在网
routing	n.	路由
home networking	n.	家庭网络

Task 1　Word bank

Directions: *Please fill in the blanks with the words listed below. You may need to change the form when necessary.*

innovation	diverse	overlap	harmonize
convergence	interaction	prevalent	bewilder
coordinate	collocate	mechanism	fragmentation

1. The society we are living in now is much more _____ than ever before.
2. He was one of the most creative and _____ engineers of his generation.
3. Millions of people want new, simplified ways of _____ with a computer.
4. His reaction had _____ her.
5. The study of sociology _____ with the study of economics.
6. In linguistics, if one word _____ with another, they often occur together.
7. This condition is more _____ in men than in women.
8. Competitors from more than a hundred countries have _____ on Sheffield for the

Games.

9. Barbara White and her mother like to listen to music together, though their tastes don't _____.

10. They spend several weeks each year undergoing intensive treatment which enables them to _____ their limbs better.

Task 2　Polysemant

In English, many words have more than one meaning. We call them polysemants. For example, "spring" could mean both "fountain" and "a season of a year". Some words may even have multiple parts of speech. For example, "sense" can be either a noun or a verb.

Please use the following words from the text to make sentences in which they were used differently.

1. coordinate
Sentence a. _____
Sentence b. _____
2. terminal
Sentence a. _____
Sentence b. _____
3. tag
Sentence a. _____
Sentence b. _____
4. overlap
Sentence a. _____
Sentence b. _____
5. interface
Sentence a. _____
Sentence b. _____

Ⅲ　Listening and Speaking

Directions: Scan the QR code on the margin and you will watch a video clip of a discussion over the IoT Standards.

Listening preparation—vocabulary

Please study the following new words which are going to appear in the listening material.

flexibility	*n.*	the quality of being adaptable or variable
critical	*adj.*	extremely important
emergence	*n.*	the process of coming into existence
accelerate	*v.*	to happen or to make sth happen faster or earlier than expected

implementation	n.	the act of accomplishing some aim or executing some order
adoption	n.	the decision to start using sth such as an idea, a plan or a name
rotate	v.	perform a job or duty on a rotating basis
outdated	adj.	no longer useful because of being old-fashioned
rigid	adj.	very strict and difficult to change
lengthy	adj.	very long, and often too long, in time or size
versus	v.	used to show that two teams or sides are against each other
vertical	adj.	having a structure in which there are top, middle and bottom levels

Task 1　Blank filling

Directions: In listening comprehension, we are often faced with crippled speech, difficult accent, occasional pause and filler words. It is crucial that we practice and master the skill of obtaining information from these speeches. Please watch the first part (0:00~1:33) of the video and try to connect the contents in it by filling in the following blanks.

1. The host's summary of previous discussion: ... the _____ of standards and how can that _____ some of decisions that your _____ would make ...

2. The host's question: Do you foresee that having the _____ to change to different systems would be a critical _____ _____ or you think that soon we would see the _____ of certain standards that would help you _____ the _____ of some of these solutions?

3. Clyde Harris's answer: It's a little bit of _____ ... Standards are going to be _____ ... At the end of the day standards are going to be a huge part of allowing _____ _____ ... but until then there is gonna be some great solutions and some great things, but when you look at mass adoption...standards need to be _____ _____ in these solution type architecture.

Task 2　Summarization

Directions: Please watch the whole part of the video, and summarize the main idea each guest wants to convey. You may watch it several times to get the correct words.

Clyde Harris	Rui Frazao	Steve Garou

Task 3　Simulating an academic talk

Direction: Please form in groups of 3~4, pick a topic related to IoT and simulate an academic talk similar to the one in the video. One of you should moderate the talk and ask questions, others should be the guests and state their opinions. Your simulation should be around 3 minutes.

Ⅳ Writing

1. Writing a research paper: body part—research methods

In academic writing, in order to solve your problem, or prove your viewpoint, you need to adopt an appropriate research method or methods. Some of the most frequently used methods include: Participant Observation, Surveys, Questionnaires, Interviews, Focus Groups, Case Study, Experiments, Secondary Data Analysis or Archival Study and Mixed Methods (combination of some of the above).

Then how do you write about your methodology? Here are some steps to follow.

Step 1: Explain your methodological approach. Begin by introducing your overall approach to the research. What research problem or question did you investigate, and what kind of data did you need to answer it? Is it a quantitative approach? Or a qualitative one? Or occasionally, both?

Step 2: Define how you collected or selected data. In this part, you should give full details of the methods you used to conduct the research; tell your readers when and where you conducted your research; outline the tools, procedures and materials you used to gather data and the criteria you used to select participants or sources.

Step 3: Describe how you analyze your data. However, you should avoid going into too much detail—you should not start presenting or discussing any of your results at this stage.

Step 4: Evaluate and justify your choice of methods. Basically, you should argue why the selected approach, instead of others, is the best for the present research. And you should also explain the limitations or weaknesses of this approach.

2. Writing a research paper: body part—result and discussion

While writing your thesis, the results and discussion sections can be both the most interesting and the most challenging sections to write. You may choose to write these sections separately, or combine them into a single chapter.

The results and discussion sections of your research paper should include the following:
- major findings;
- comparison with previous studies;
- limitations of your work;
- casual arguments;
- speculations;
- deductive arguments.

(http://www.scribbr.com/dissertation/methodology)

Task 1 Conduct a questionnaire

Directions: Questionnaire is one of the research methods. Please conduct a questionnaire with 6 to 10 questions in the aim of finding out people's opinions on the standardization of IoT. Please write your tentative questions below.

___

Task 2 Writing a result and discussion report

Directions: Please choose a small group of people to fill in the questionnaire you have designed in Task 1. Then gather their responses, analyze them and write a report on the result and analysis. (100~200 words)

___

KEY

Ⅰ Reading

Task 1

1. reality, coordinated, item labeling, process control, wireless technology, network interconnection

2. harmonizing specifications, interoperable global standards

3. identification, transferred, devices, detect, control, networking, connected

Task 2

1. A 2. B 3. D 4. A 5. C

Chapter 1　IoT Technology Fundamentals

Ⅱ　Vocabulary

Task 1

1. diverse　2. innovative　3. interacting　4. bewilder　5. overlaps
6. collocates　7. prevalent　8. converged　9. harmonize　10. coordinate

Task 2

1. coordinate
a. We need to develop a coordinated approach to the problem.
b. Can you give me your coordinates so I can find you?
2. terminal
a. Carl sits at a computer terminal 40 hours a week.
b. A terminal illness is often difficult to be cured and will cause death.
3. tag
a. A hospital is to fit newborn babies with electronic tags to stop kidnappers.
b. I let him tag along because he had not been too well recently.
4. overlap
a. A fish's scales overlap each other.
b. There was no overlap between their proposals.
5. interface
a. The new system interface is more user-friendly.
b. The committee calls for standardized interfaces among equipments so that products made by different companies could be compatible.

Ⅲ　Listening and Speaking

Task 1

1. lack, impact, clients
2. flexibility, success factor, emergence, accelerate, implementation
3. both, critical, mass adoption, in place

Task 2

Clyde Harris: Standards are going to be critical for mass adoption in future. But before standards are set, there should also be some other more flexible solutions that allow the different layers to interoperate.

Rui Frazao: To be flexible, we should have numbers of standards that we shouldn't be too rigid about, and we should leave out unnecessary and lengthy standards.

Steve Garou: The interoperability of different layers is good for future business growth.

Script for Listening

Host: The discussion it was mentioned that the lack of standards and how can that impact some of the decisions that your clients would make, so do you foresee that having the flexibility to change to different systems will be a critical success factor or you think that soon we can see the emergence of certain standards that will help accelerate the implementation of some of these solutions.

Clyde: It's a little bit of both. But I would in our industry where there's banking payments, connectivity, authentication into buildings and government access standards. Standards are going to be critical for mass adoption of at least in some sense but always before standards comes this kind of stepping stone of new technology creations and solutions that are defined from within the companies that developed them. But at the end of the day I think standards are going to be a huge part of allowing mass adoption in an affordable way from a business perspective. But until then there's gonna be some great solutions and some great things but when you look at mass adoption like everything else the computer industry, the cell phone industry, standards need to be in place in these solution type architectures.

Rui Frazao: Yeah well in the carrier side when you talk standard you need to be careful, because they over rotate on standards and that's probably one of the reasons why their architectures are so outdated. But but on the other side you see today carriers are talking about open source and that's a way of them to express that okay there's an open community discussing we should interface with each other and we move away from our rigid model of standards. So definitely we need to have a number of standards that allow the different layers to interoperate. But we should be careful the carriers I think I've been careful that we're not following the model of the telecoms industry with this under the old type of standards that tends to be very complex and lengthy. So there's a need for a speedy process, so we need to focus on what is relevant. I mean those standards and leave out what probably can be handled within a certain layer of technology.

Clyde: Yeah once you get to the standard, the flexibility of changing the standard would be great.

Rui Frazao: Exactly.

Steve Garou: The word that we actually use is interoperability and the faster that we can get components and software and services to use interoperable pieces and know how they work together, the better it is for our business because then we can sell more. Until that happens then we're kind of having to talk about Best of Breed we're kind of having to encourage one versus the other but it's that interoperability I think more so than standards that'll really drive growth.

Clyde: Especially in this industry where if you look at the circle of the ecosystem and like the different verticals you were talking about and plugging this into that and this into that, it's just natural that you're gonna drive costs down and increase applications and only with some weighing that way, and it's coming. But there's so many things you can connect. It's confusing.

The Applications of the IoT

- How the Internet of Things Can Prepare Cities for Natural Disasters
- IoT in Action in Healthcare
- Flying Smarter — The Smart Airport and the Internet of Things
- Industrial Internet of Things
- The IoT Data Opportunity for Logistics Companies Is here

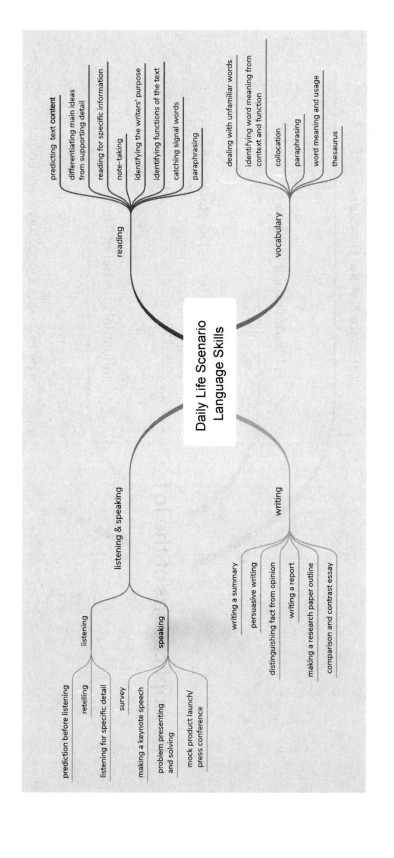

Chapter 2 The Applications of the IoT

This chapter provides 5 articles on some applications of IoT technologies. You will be put in daily life scenario, in which you will learn how to read with effective strategies, compose practical writing, and communicate in different common situations.

Unit 6　How the Internet of Things Can Prepare Cities for Natural Disasters

1　When a disaster strikes, federal, state, and local governments need a coordinated strategy, accessible data, and a skilled workforce to manage the response. Natural disasters such as hurricanes, tornadoes, and floods have devastating effects on communities across our country. Since 1980 the U. S. has sustained more than 200 weather and climate disasters, with cumulative costs exceeding $1.1 trillion.

2　Government agencies should consider leveraging the Internet of Things (IoT) and other web-driven technologies to obtain timely and accurate data that can better inform decisions and actions. Using the most current technology could help them more efficiently and safely address these costly disasters. However, this type of progress will require more than just employing the IoT to improve emergency preparedness and response; response teams have to be ready to receive, interpret, and take action on the data.

---

tornado /tɔːˈneɪdəʊ/ *n.*
a violent wind storm consisting of a tall column of air which spins around very fast and causes a lot of damage

devastating /ˈdevəˌsteɪtɪŋ/ *adj.*
very harmful or damaging

cumulative /ˈkjuːmjʊlətɪv/ *adj.*
If a series of events have a cumulative effect, each event makes the effect greater.

leverage /ˈliːvərɪdʒ/ *v.*
to get as much advantage or profit as possible from sth that you have

preparedness /prɪˈpeərɪdnɪs/ *n.*
the state of being ready for something to happen, especially for war or a disaster

Gathering data before a disaster strikes

3   Today, disaster responders gain reliable, timely information only when they reach an emergency zone and take stock of the situation. In the case of hurricanes and major weather events, physical and technical roadblocks often prevent response teams from obtaining critical data to track damages, prioritize response needs, and keep the public informed so that people know how to stay safe. Ineffective communication channels, overburdened response systems, satellite disruptions, and Internet blackouts further impede people from getting the help they need.

4   That's where the value of IoT sensors that collect data and systematically broadcast signals from emergency areas comes into play. These sensors can relay information about their surroundings directly to government agencies and emergency teams. For example, sensors can measure temperature, water quality, pressure, level, smoke, and humidity, to name just a few uses. In the case of wildfires, sensors can detect how far and how fast is the fire spreading. For hurricanes or tsunamis, sensors can monitor water levels to send alerts at the first sign of flooding. Sensors can also be used to detect the presence of harmful gases or chemicals emanating from a storage tank, factory, or plant in the path of destruction. These devices can be critical for urgent decisions like whether to evacuate an area at risk of flooding, or how to guide residents to the safest exit routes ahead of an emergency.

5   In practice, this starts with establishing systems that connect local data to government responders. Technical teams could deploy sensors that send web-linked data to a digital command center that government officials can access remotely while at the scene. **Drones** could surveil disaster areas during the search-and-rescue phase and then move to data collection to support the recovery effort once the immediate danger has passed.

6   In order to optimize effectiveness, agencies should place web-linked sensors on physical assets such as levees, bridges, and **utility poles** to monitor risk factors such as rising water levels in low-lying areas and to alert authorities when there's an issue with critical infrastructure. In areas vulnerable to flooding, for example, response teams should arrange sensors in various

---

take stock of
v. to make an itemized list or record of the resources or goods available, in stock, or in one's possession

prioritize /praɪˈɒrɪtaɪz/ *v.*
treat it as more important than other things

overburdened /ˌəʊvəˈbɜːdənd/ *adj.*
If a system or organization is overburdened, it has too many people or things.

blackout /ˈblækaʊt/ *n.*
a period of time during a war in which towns and buildings are made dark so that they cannot be seen by enemy planes

impede /ɪmˈpiːd/ *v.*
(formal) If you impede someone or something, you make their movement, development, or progress difficult.

relay /rɪˈleɪ/ *v.*
(formal) If you relay something that has been said to you, you repeat it to another person

humidity /hjuːˈmɪdɪtɪ/ *n.*
the air being very heavy and damp

tsunami /tsʊˈnæmɪ/ *n.*
a very large wave, often caused by an earthquake

emanate /ˈɛməneɪt/ *v.*
If a quality emanates from you, or if you emanate a quality, you give people a strong sense that you have that quality.

evacuate /ɪˈvækjʊeɪt/ *v.*
To evacuate someone means to send them to a place of safety, away from a dangerous building, town, or area.

drone /drəʊn/ *n.*
(informal) an aircraft that does not have a pilot, but is operated by radio

surveil /sɜːˈveɪl/ *v.*
to observe closely the activities of (a person or group)

levee /ˈlɛvɪ/ *n.*
a raised bank alongside a river

utility pole A utility pole is a tall pole with telephone or electrical wires attached to it.

low-lying /ˈləʊˈlaɪɪŋ/ *adj.*
land at, near, or below sea level

locations so that one device going down won't take down the entire network. Establishing a stream of data from sensors in at-risk areas can also help <u>pinpoint</u> and prioritize which neighborhoods need to be reached first.

7　Response teams can gain an even clearer picture of the emergency situation if the sensor data is combined with Census-verified <u>demographic</u> and relevant third-party data. Increased socioeconomic and demographic data would be useful to inform <u>outreach</u> <u>tactics</u>, for example in a community where people speak many different languages. Analytics-backed information would enable local, state, and national teams to **geotarget** messaging to neighborhoods at most risk—a neighborhood with high <u>concentrations</u> of elderly populations who might not have access to transportation, for example.

Connecting people and information during a disaster

8　In order to respond with precision, government agencies and emergency response teams should establish communication systems between the mobile devices of an at-risk area's residents and IoT sensors in the field. Doing so can help facilitate and <u>expedite</u> a local response during the disaster. The system should respond to <u>incoming</u> information based on data it receives from the IoT sensors and signals from citizens' mobile devices. For example, if a citizen messages a public emergency text line to ask where to go to avoid local flooding, the system could provide a recommendation based on water level data it receives from <u>deployed</u> sensors. A data-backed automated response can ensure information reaches the people who need it most. This data should be collected centrally, monitored regularly by response officials, and proactively used to inform automated alerts that are broadcasted to citizens' mobile devices within a certain <u>radius</u> of the hazard area.

9　Response teams can also use the sensor data for coordination, analytics, outreach strategies, and on-the-ground tactics. These actions will vary from case by case. In the case of a food stamp program, government officials could use the information to decide ① how and when to reach out to the affected population, ② where to set up temporary benefit distribution centers, because the primary centers (supermarkets, convenience stores, and so on) may not be functional, and ③ how to ensure benefits

---

pinpoint /ˈpɪnˌpɔɪnt/ v.
If you pinpoint sth or its position, you discover or show exactly where it is.

demographic /ˌdɛməˈɡræfɪk/ n.
statistics relating to the people who live in a place

outreach /ˈaʊtriːtʃ/ n.
the activity of an organization that provides a service or advice to people in the community, especially those who cannot or are unlikely to come to an office, a hospital, etc. for help

tactic /ˈtæktɪk/ n.
the particular method you use to achieve sth

concentration /ˌkɒnsənˈtreɪʃən/ n.
a large amount of it or large numbers of it in a small area

expedite /ˈɛkspɪdaɪt/ v.
(formal) to make a process happen more quickly

incoming /ˈɪnˌkʌmɪŋ/ adj.
arriving somewhere, or being received

deploy /dɪˈplɔɪ/ v.
(formal) to use sth effectively

radius /ˈreɪdɪəs/ n.
a straight line between the centre of a circle and any point on its outer edge

are distributed correctly.

10   Emergency response organizations must also know which communication channels work best to reach the affected citizens. For instance, if the at-risk population is predominantly Spanish-speaking, then preparedness messaging should be done in Spanish. When dealing with an elderly population, the outreach can be done through television, newspapers, and radio rather than tech-driven channels like text alerts and apps. This targeted communication is a shift from the conventional "one size fits all" approach. Agencies will need both a thorough change management process to describe the benefits and efficacy of the IoT-based approach and appropriate training in how to adopt it.

11   We're seeing the early stages of IoT-based response happen already with the **Department of Homeland Security（DHS）'s Consolidated Asset Portfolio and Sustainability Information System （CAPSIS）** system. Responders to the hurricanes in Texas and Puerto Rico and the wildfires in California used field data streaming to CAPSIS at DHS headquarters to take stock of the damage. At the state-level, Wyoming's Department of Transportation has rolled out a pilot program to use vehicle-to-vehicle, vehicle-to-infrastructure, and infrastructure-to-vehicle connectivity to improve monitoring and reporting of road conditions to drivers along I-80.

12   Timing is everything in a disaster situation. By incorporating IoT data into emergency response plans, public sector agencies and responders can use real-time information to make plans and reach the citizens who need help.

Word count:1,066

Source: Tremaine K, Tuberson K. How the Internet of Things can prepare cities for natural disasters[EB/OL].(2017-12-01)[2019-05-15]. https://hbr.org/2017/12/how-the-internet-of-things-can-prepare-cities-for-natural-disasters.

---

consolidate /kənˈsɒlɪdeɪt/ v.
to make a position of power or success stronger so that it is more likely to continue

portfolio /pɔːtˈfəʊlɪəʊ/ n.
the range of products or services offered by a particular company or organization

---

Ⅰ Reading

Reading for better understanding of the main idea: differentiating between main ideas and supporting details.

In previous learning, you were introduced to read for a purpose, i.e., concentrating on the texts or parts of texts that are relevant to your reading purpose. Your reading purpose now is to identify main ideas.

Key reading skills: differentiating main ideas from supporting details.

In everyday reading and especially academic context, it is important to be able to extract the main ideas from a text, particularly if the text contains complex ideas and a lot of supporting points. A paragraph generally contains one main idea and may have several supporting details. You first need to identify the main points and extract the key information. You can then decide which of the supporting details are also relevant to your reading purpose. This key reading skill is particularly useful for increasing your reading speed and understanding.

Task 1

Directions: Re-read the text, study the information in sentences 1~6 below, find the relevant sections of the text and select three pieces of information which are main ideas.

a. Response teams can also use the sensor data for coordination, analytics, outreach strategies, and on-the-ground tactics.

b. Sensors can also be used to detect the presence of harmful gases or chemicals emanating from a storage tank, factory, or plant in the path of destruction.

c. Emergency response organizations must also know which communication channels work best to reach the affected citizens.

d. These devices can be critical for urgent decisions like whether to evacuate an area at risk of flooding, or how to guide residents to the safest exit routes ahead of an emergency.

e. A data-backed automated response can ensure information reaches the people who need it most.

f. That's where the value of IoT sensors that collect data and systematically broadcast signals from emergency areas comes into play.

In previous learning, you were introduced to ways of taking notes by creating a mind map. Your task below is to practice different note taking strategies in reading and apply accordingly to record main ideas.

Different people organize their notes in different ways. Some students write linear notes, starting at the top of the page and working down, while other students prefer to use mind maps. The best solution may be to use different ways of taking notes for different types of articles.

Linear notes are arranged so that the ideas are set out one after the other.

Task 2

Directions: Linear notes. You can start with identifying the thesis statement (the central idea) by reading the first two paragraphs and find out the topic sentences for each paragraph.

```
                        Linear notes
Thesis Statement: (a)
                (b)
Topic Sentence 1
Topic Sentence 2
Topic Sentence 3
...
```

Directions: Mind maps. You can also create a mind map by filling in the blanks of the example below.

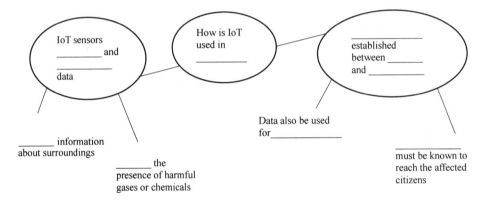

Ⅱ Vocabulary

Key words and expressions

tornado	devastating	cumulative
leverage	preparedness	take stock of
prioritize	overburdened	blackout
impede	relay	humidity
tsunami	emanate	evacuate
drone	surveil	levee
low-lying	pinpoint	demographic
outreach	tactic	concentration
expedite	incoming	deploy
radius	consolidate	portfolio

Terminology

drone	n.	无人机
utility pole	n.	电线杆
geotarget	n.	地理定位
Department of Homeland Security (DHS)	n.	美国国土安全部
Consolidated Asset Portfolio and Sustainability Information System (CAPSIS)	n.	强化资产登记和可持续信息系统

Task 1

Dealing with unfamiliar words: identifying word meaning from context and function.

Directions: Please work out meanings of the words in the table. Follow the procedure below.

a. Locate each word in the text and identify its word class. Complete column 2 of the table.

b. Look for clues in the surrounding context to help you guess the meaning of each word. Complete column 3.

c. Compare your answers with another student's. Work together to think of a possible synonym (word or short phrase) for each word and complete column 4. Remember that the synonym must be the same word class and be able to replace the original word in the context.

d. Check meanings and synonyms in a dictionary, if necessary.

Word in text	Word class	Meaning in context	Possible synonym
effectiveness	noun	the degree to which something is successful	efficacy; success
vulnerable			
stream			
concentration			
tactics			
functional			
predominately			
pilot			
timing			

Task 2

Directions: Analyze the article in detail and explain in your own words.

Why do you think some phrases in the text have been underlined?

a. Discuss the meaning of each phrase with another student, or work alone.

b. Complete the table below by explaining, in your own words, the meaning of each phrase.

Emergency response organizations must also know which communication channels work best to reach the affected citizens. For instance, if the at-risk population is predominantly Spanish-speaking, then preparedness messaging should be done in Spanish. When dealing with an elderly population, the outreach can be done through television, newspapers, and radio rather than tech-driven channels like text alerts and apps. This targeted communication is a shift from the conventional "one size fits all" approach. Agencies will need both a thorough change management process to describe the benefits and efficacy of the IoT-based approach and appropriate training in how to adopt it.

Phrase	Paraphrase
affected citizens	
at-risk population	
preparedness messaging	
tech-driven channels	
targeted communication	
a shift from the conventional "one size fits all" approach	

Ⅲ Listening and Speaking

Directions: Scan the QR code on the margin and you will listen to a piece of news about the introduction of IoT, preparing yourself for the following tasks.

Task 1

Prediction before listening:

What aspects of IoT will you introduce to people who are unfamiliar with it? Imagine explaining its concept and application to your father, mother and other elder relatives, what questions will they ask, and what are you going to say?

Task 2

Retelling after listening:

Listen to the news twice. Take notes for the second time about the questions asked by the hostess and answers offered by the expert. Retell the whole interview and cross-checking with your classmates.

When you practice retelling, time yourself in 3 minutes, and produce your speech based on an organized, logical processing and summarizing of the notes.

Task 3

Reflection after listening:

Are you satisfied with the answer provided by the expert? What example would you use if you are going to tell someone about IoT? How do you understand "more efficiency and better understanding like a layer of data on top of the world we're living now"?

Ⅳ Writing

Task 1 Write a summary

Directions: Read the introduction (the first two paragraphs) and the first section (all paragraphs following the first subheading), and do not refer to your previous notes. Your teacher will set a time limit.

When developing your ability in summarizing, it is advisable to practice using one particular approach. The NOW approach is one way of doing this.

Note: making notes on the main points

Organize: organizing these notes

Write: writing up the notes in an appropriate format

You will be able to compare your notes and summary with examples supplied by the teacher.

Write your notes

Identify your note-making purpose, e. g. to summarize the whole text (a "global" summary), or specific parts of the text that are relevant to a particular purpose (a "selective" summary).

Write down everything relevant to the main ideas in linear form or as a mind map, using your own words where possible. Check the original text to make sure that you have not missed out any important information.

For example:

Introduction	Data gathering	Data connecting

Key skills: You can often save time by making notes after you have finished reading. If you make notes as you read, it may slow down your reading and interrupt your concentration on the text.

After writing down your notes, you can discuss with your classmates, and compare your notes with the model supplied by your teacher.

a. What do you find most interesting or effective about the notes provided by your teacher?

b. Do you think your own notes are effective?

Organize the notes

Study the notes and decide in which order you wish to put the information when you write the summary. This will depend on why you are making the notes. Decide how you want to prioritize the information. You may want to reorganize the ideas and information.

Write your summary

A summary is a shortened passage, which retains the essential information of the original. It is fairly brief restatement— in your own words—of the contents of a passage.

Key skills: When you grasped the central idea, use paraphrasing to condense the meaning. You can either use synonyms or synonymous phrases, change the structure of sentences, or combine sentences. It is also essential to connect all the topic sentences in the article with major supporting ideas.

Edit your first draft; make sure you have included all the relevant information and checked the accuracy of your grammar, vocabulary and spelling. Write out a second draft, if necessary.

KEY

Ⅰ Reading

Task 1

a. c. f.

Task 2

Linear notes

Thesis Statement: (a) Government agencies should consider leveraging the IoT in disasters to better address them.

(b) The progress with IoT employed will require response teams to cope with the data in every way.

Topic Sentence 1
Topic Sentence 2
Topic Sentence 3
…

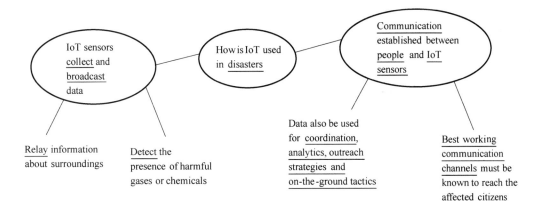

II Vocabulary

Task 1

Word in text	Word class	Meaning in context	Possible synonym
effectiveness	noun	the degree to which something is successful	efficacy; success
vulnerable	adjective	easily threatened by; exposed to possible dangers	liable; defenseless
stream	noun	an ongoing flow; a series of	series; chain
concentration	noun	relative amount of things gathered in a volume or space	density; cluster
tactics	noun	actions or moves carefully planned to achieve a purpose	strategies; scheme
functional	adjective	working in normal conditions	working; operating
predominately	adverb	for the most part	mainly; largely
pilot	adjective	done as an experiment or test before introduced more widely	trial; prior
timing	noun	controlling of the time	scheduling

Task 2

Phrase	Paraphrase
affected citizens	citizens threatened by natural disasters
at-risk population	citizens threatened by natural disasters/people whose lives are at risk in disasters
preparedness messaging	the message sending as a note to prepare people for the disaster
tech-driven channels	methods that are mainly based on high technology (such as apps)
targeted communication	communication that is aimed to serve particular groups (such as the elderly)
a shift from the conventional "one size fits all" approach	a change different from the traditional method that is applied universally without any concern for adjustments

Ⅳ Writing

Task1

Introduction	Data gathering	Data connecting
Government should consider leveraging IoT to reach better decisions, which will also require response teams to collaborate timely.	IoT sensors can collect and systematically broadcast data, reaching quickly to government responders, and so agencies need to place the sensors wisely and consider the combination of data from different sources.	Communication systems between government and the at-risk area's residents are vital to the quick response, which will require an optimization of the communication channel and good timing.

Script for Listening

As we rely more and more on Internet-connected devices, experts say, this year we are going to see more advances, especially in your home, by colleague Green spoken with CNET senior producer Dan Patterson about the Internet of Things.

D: It sounds like a lot of jargon, but if you think about processing power and connectivity and we just mean computer chips in the Internet embedded into everything. So we've heard a lot of jokes and commentaries about your refrigerator or your washer drier. That sounds funny but if you think about it: your washing machine had scale. If it would understand the water flow, this saves us tons of energy on the long run. So the Internet of things is really just connectivity and processing power embedded into the world around us, and that's, this is also the industrial world, it is our city, so we are going to hear a lot about smart cities, about smart cars and about smart home. Because when everything is connected, all of a sudden we have data that allows us to really understand the world in ways that we were never able to before.

G: Dan you were just off your heels from your trip to Davos, last week, and you were saying that there's this sort of concept that fourth Industrial Revolution with the Internet of Things, explain that.

D: Yeah. The fourth Industrial Revolution, again it sounds like a lot of jargon. But what this really means is that data is changing everything, so really the Internet of Things if we embed processing power into common objects, it means that we can extract data and information. Data and information also feed artificial intelligence. Artificial intelligence is then used in algorithms to automate a lot of things that are manual, that we use human processing power for. So it's kind of this weird, virtuous circle of technologies that are getting smaller and more powerful, and when I say it far more again it's going to sound like jargon. But I mean instead of the slap of black plastic and steel in the pockets, it's really the world around us.

G: So when you talk about it will affect everyday life. For instance, the milk in my fridge starts to run out. Are we talking about somehow it will automatically be refilled or is that what we are talking about?

D: Yeah that's one real basic understanding of the Internet of Things, but if we think about traffic flow and traffic patterns. If we have sensors in every traffic light around us, and sensors in our car and we have what's called vehicle-to-vehicle communication. We don't really need to be looking at this traffic maps and then Google tells us what's red so we gotta slow down going over the Brooklyn Bridge. It means that our cars are communicated with each other, and traffic flows seamlessly as opposed to stop and go. Really what we are talking about, with the Internet of Things, is more efficiency [SIC] and better understanding like a layer of data on top of the world we're living now.

G: Dan Patterson. Thank you Dan.

Unit 7   IoT in Action in Healthcare

1   The IoT plays a significant role in a broad range of healthcare applications, from managing <u>chronic</u> diseases at one end of the <u>spectrum</u> to preventing disease at the other. Here are some examples of how its potential is already playing out.

2   Clinical care: Hospitalized patients whose physiological status requires close attention can be constantly monitored using IoT-driven, noninvasive monitoring. This type of solution employs sensors to collect comprehensive physiological information and uses **gateways** and the cloud to analyze and store the information and then send the analyzed data wirelessly to caregivers for further analysis and review. It replaces the process of having a health professional come by at regular intervals to check the patient's vital signs, instead providing a continuous automated flow of information. In this way, it simultaneously improves the quality of care through constant attention and lowers the cost of care by <u>eliminating</u> the need for a caregiver to actively engage in data collection and analysis.

3   An example of this type of system is the **Masimo Radical-7**, a health monitor for clinical environments that collects patient data and wirelessly transmits for ongoing display or for notification purposes. The results provide a complete, detailed picture of patient status for <u>clinicians</u> to review wherever they may be. The monitor <u>incorporates</u> Freescale technology in the form of an **i. MX** applications processor with enhanced graphics capabilities that enables the extremely **high-resolution** display of information, as well as a touch-based user interface that makes the technology easy to use.

4   Remote monitoring: There are people all over the world whose health may suffer because they don't have ready access to effective health monitoring. But small, powerful wireless solutions connected through the IoT are now making it possible for monitoring to come to these patients instead of vice-versa. These solutions can be used to securely capture patient health data from a variety of sensors, apply complex <u>algorithms</u> to analyze the data and then share it through wireless connectivity with medical professionals who can make <u>appropriate</u> health

---

chronic /ˈkrɒnɪk/ *adj.*
(especially of a disease) lasting for a long time; difficult to cure or get rid of

spectrum /ˈspektrəm/ *n.*
a complete or wide range of related qualities, ideas, etc.

gateway /ˈɡeɪtˌweɪ/ *n.*
a device that connects two computer networks that cannot be connected in any other way

eliminate /ɪˈlɪmɪˌneɪt/ *v.*
To eliminate sth, especially sth you do not want or need, means to remove it completely.

clinician /klɪˈnɪʃən/ *n.*
a doctor, psychologist, etc. who has direct contact with patients

incorporate /ɪnˈkɔːpəˌreɪt/ *v.*
to include sth so that it forms a part of sth

algorithm /ˈælɡəˌrɪðəm/ *n.*
a set of rules that must be followed when solving a particular problem

appropriate /əˈproʊprɪɪt/ *adj.*
Something that is suitable or acceptable for a particular situation

recommendations.

5   As a result, patients with chronic diseases may be less likely to develop complications, and acute complications may be diagnosed earlier than they would be otherwise. For example, patients suffering from cardiovascular diseases who are being treated with digitalis could be monitored around the clock to prevent drug intoxication. Arrhythmias that are randomly seen on an EKG could be easily detected, and **EKG** data indicating heart hypoxemia could lead to faster detection of cardiac issues. The data collected may also enable a more preventive approach to healthcare by providing information for people to make healthier choices.

6   An example of an enabling technology for remote monitoring is the Freescale Home Health Hub reference platform, which is built on Freescale i. MX applications processing technology and tightly integrates key capabilities—such as wireless connectivity and power management—in the telehealth gateway that enables collection and sharing of physiological information. The hub captures patient data from a variety of sensors and securely stores it in the cloud, where it can be accessed by those engaged in the patient's care. Data aggregation devices like this will soon become commonplace and will not only collect healthcare data but also manage other sensor networks within the home. Freescale's second-generation gateway manages data from smart energy, consumer electronics, home automation and security systems—in addition to healthcare.

7   Early intervention/prevention: Healthy, active people can also benefit from IoT-driven monitoring of their daily activities and well-being. A senior living alone, for example, may want to have a monitoring device that can detect a fall or other interruption in everyday activity and report it to emergency responders or family members. For that matter, an active athlete such as a hiker or biker could benefit from such a solution at any age, particularly if it's available as a piece of wearable technology.

8   Freescale technology has been incorporated into some solutions of this type. The Sonamba daily monitoring solution, aimed at the senior population, uses strategically placed sensors to monitor daily activities and report anomalies to care providers or family members via cell phone. Freescale provides applications

---

complication /ˌkɒmplɪˈkeɪʃən/ *n.*
a new problem or illness that makes treatment of a previous one more complicated or difficult

acute /əˈkjuːt/ *adj.*
You can use acute to indicate that an undesirable situation or feeling is very severe or intense.

cardiovascular /ˌkɑːdɪəʊˈvæskjʊlə/ *adj.*
connected with the heart and the blood vessels

digitalis /ˌdɪdʒɪˈteɪlɪs/ *n.*
a drug made from the foxglove plant, that helps the heart muscle to work

around the clock day and night

intoxication /ɪnˌtɒksɪˈkeɪʃən/ *n.*
the state of being drunk

arrhythmia /əˈrɪðmɪə/ *n.*
any variation from the normal rhythm in the heartbeat

hypoxemia /ˌhaɪpɒkˈsiːmɪə/ *n.*
a lower than normal amount of oxygen in the blood

cardiac /ˈkɑːdɪæk/ *adj.*
connected with the heart or heart disease

hub /hʌb/ *n.*
the central and most important part of a particular place or activity

aggregation /ˌægrɪˈgeɪʃən/ *n.*
the formation of a number of things into a cluster

commonplace /ˈkɒmənpleɪs/ *adj.*
done very often, or existing in many places

intervention /ˌɪntəˈvɛnʃən/ *n.*
the act of becoming involved in an argument, fight, or other difficult situation in order to change what happens

anomaly /əˈnɒməlɪ/ *n.*
a thing, situation, etc. that is different from what is normal or expected

processing and ZigBee-based wireless connectivity for Sonamba. Freescale technology is also embedded in the Numera Libris mobile personal health gateway, which is designed to detect falls and provide the ability to manage one's health at home or away.

9 These are just a few examples of IoT-based healthcare solutions, and many more are emerging. But as one reporter has noted, "The real vision for the future is that these various smaller applications will <u>converge</u> to form a whole... Imagine if you are a relative of a patient who forgot his medicine. You receive the alert, are able to know his location, check his vital signs remotely to see if he is falling ill, then be informed by your car's <u>navigation</u> system which hospital has the most free beds, the clearest traffic route to get there and even where you can park."

10 The successful use of the IoT in the preceding healthcare examples relies on several enabling technologies. Smart sensors, which combine a sensor and a **microcontroller**, make it possible to harness the power of the IoT for healthcare by accurately measuring, monitoring and analyzing a variety of health status indicators. These can include basic vital signs such as heart rate and blood pressure, as well as levels of <u>glucose</u> or oxygen <u>saturation</u> in the blood. Smart sensors can even be incorporated into pill bottles and connected to the network to indicate whether a patient has taken a scheduled dose of medication. For smart sensors to work effectively, the microcontroller components must incorporate several essential capabilities.

11 Low-power operation is essential to keeping device footprint small and extending battery life, characteristics that help make IoT devices as usable as possible. Freescale, which has long offered low-power processing, is working now to enable completely **battery-free** devices that utilize energy harvesting techniques through the use of ultra-low-power **DC-DC** converters.

12 Integrated precision-analog capabilities make it possible for sensors to achieve high accuracy at a low cost. Freescale offers this enabling technology within microcontrollers which contain <u>analog</u> components, such as high-resolution analog-to-digital converters (**ADCs**) and low-power **Op-Amps.**

13 Graphical user interfaces (GUIs) improve usability by enabling display devices to deliver a great deal of information in vivid detail and by making it easy to access that information.

---

converge /kənˈvɜːdʒ/ *v.*
to move towards a place from different directions and meet

navigation /ˌnævɪˈɡeɪʃən/ *n.*
the skill or the process of planning a route for a ship or other vehicle and taking it there

glucose /ˈɡluːkəʊz/ *n.*
a type of sugar that is found in fruit and is easily changed into energy by the human body

saturation /ˌsætʃəˈreɪʃən/ *n.*
the degree to which sth is absorbed in sth else, expressed as a percentage of the greatest possible

analog /ˈænəlɒɡ/ *adj.*
using a continuously changing range of physical quantities to measure or store data

Freescale's i. MX applications processors with high graphics-processing performance support advanced GUI development.

14 Gateways are the information hubs that collect sensor data, analyze it and then communicate it to the cloud via wide area network (**WAN**) technologies. Gateways can be designed for clinical or home settings; in the latter, they may be part of larger connectivity resource that also manages energy, entertainment and other systems in the home. The Freescale Home Health Hub reference platform includes a gateway component. Medical device designers can also use the platform to create remote-access devices for remote monitoring.

Word count: 1,058

Source: Adapted from Niewolny D. How the Internet of Things is revolutionizing healthcare[EB/OL]. (2013-10)[2019-05-15]. https://www.nxp.com/files-static/corporate/doc/white_paper/IOTREVHEALCARWP.pdf.

I Reading

Predicting text content

Predicting the content of a text will help you read with more speed and fluency. It may also help you identify the writer's purpose and recognize "new" knowledge.

Task 1

Directions: Think about the title: "IoT in Action in Healthcare".

Discuss what you think healthcare includes, and why is there IoT involved in healthcare?

Add four more possible aspects where IoT can be applied in healthcare:

1. telemedicine solutions that help people to easily get prescribed drugs
2. _____
3. _____
4. _____
5. _____

What do you think is Freescale?

_____
_____

Directions: Write down from your own experience one aspect that IoT is used in connecting things in the hospital. Or you can detect problems you notice that need to be solved with IoT.

_____
_____

Directions: While you are reading, you can also time yourself. Note the time before you begin reading and note it again when you have finished. There are 1,058 words in the text.

Predicting involves using the knowledge you already have about a topic to help you understand a text you are going to read and quickly prepare yourself for the upcoming knowledge.

Now read to see if your ideas were the same as those in the text.

Directions: Tick a statement that most closely reflects the writer's viewpoint.

☐1. The cost of regular visiting to health professionals are not as cheap and quality guaranteed as using an ongoing automated flow of information from IoT devices.

☐2. Traditional healthcare solutions cannot provide people with the care and service they need.

☐3. It is very likely that soon in the future, most smart devices could read people's health information and other data to be used for their diagnoses.

☐4. Healthcare is also needed for people that are not currently suffering from any disease.

Identifying the writer's purpose

Task 2

Directions: Please answer the following questions.

1. The writer uses a range of examples to illustrate his writing purpose. What do you consider is the writer's main purpose? Choose one of the following options. (　　)

A. To outline recent developments in IoT applied in healthcare

B. To illustrate the differences between IoT in healthcare and traditional healthcare

C. To compare opinions about what IoT in healthcare is

D. To describe the history of IoT in healthcare

2. Explain your choice, and you can discuss the writer's purpose of writing each paragraph.

_____
_____

Ⅱ Vocabulary

Key words and expressions

chronic	spectrum	gateway
eliminate	clinician	incorporate
algorithm	appropriate	complication
acute	cardiovascular	digitalis
around the clock	intoxication	arrhythmia
hypoxemia	cardiac	hub
aggregation	commonplace	intervention
anomaly	converge	navigation
glucose	saturation	analog

Chapter 2 The Applications of the IoT | 87

Terminology

gateway	n.	网关
Masimo Radical-7	n.	迈心诺 Radical-7 血氧监测仪
i.MX	n.	一种图形处理器
high-resolution	adj.	高分辨率的
EKG(electrocardiogram)	n.	心电图
microcontroller	n.	微控制器
DC-DC	n.	DC-DC 转换器是一种在直流电路中将一个电压值的电能变为另一个电压值的电能的装置
ADC(analog-to-digital converter)	n.	模/数转换器(模数转换器)
Op-Amp(operation amplifier)	n.	运算放大器
GUI(graphical user interface)	n.	图形用户界面
WAN(wide area network)	n.	广域网

Task 1

Directions: Look at the words and phrases in the table below and find words and phrases in the text with a similar meaning.

Definition	Line number	Word or phrase	Word class
a. range, in terms of its position on a scale of two extremes (para. 1)			
b. all inclusive, complete (para. 2)			
c. continuing, still in progress (para. 3)			
d. designed to keep something undesired, such as illness from occurring (para. 5)			
e. make things possible and give powers (para. 6)			
f. the state of being comfortable, healthy, or happy (para. 7)			
h. previous, former (para. 10)			

Task 2 Word bank

Directions: Please fill in the blanks with the words listed below. You may need to change the form when necessary.

complication	intervention	converge
intoxication	appropriate	hub
aggregation	navigation	spectrum
chronic	saturation	incorporate

1. Then there is Shanghai's lack of experienced financial professionals— the lifeblood of any financial _____.

2. The bad weather added a further _____ to our journey.

3. An _____ of fragments is the only kind of whole we now have.

4. The rest of the business owners fall somewhere between those two ends of the _____.

5. That is, we want to have instruments that control the variables without requiring _____ from the operator.

6. They learn how to diagnose and manage common illnesses like colds, or _____ problems such as diabetes and heart disease.

7. The book was written in a style _____ to the age of the children.

8. What is odd in this scenario is that up until a month ago, no man in Libya publicly displayed _____ —not least because alcohol is illegal here.

9. To implement this function, it must be a complete stable network, namely, its all output tracks must _____ at a stable equilibrium point.

10. There's more than one way to learn, so be sure to _____ different methods in your training.

Ⅲ Listening and Speaking

Task 1 Making a keynote speech as a product launch release

Learn from the language style

Watch the video by scanning the QR code. Watch and summarize orally what are the features of this product, and what about the language the speaker used to describe it. You are encouraged to exchange ideas with your classmates.

Watch the video again and select a favorite part with a length about 3 minutes long. Carefully watch the favored part, take notes, and then mute the video to perform the speaker's keynote by yourself.

Video Clip

Release your own product

Create your own keynote speech on a favorite technological product of yours. You need to: create slides with clear, focused information; summarize the 3 most striking features of the product and illustrate with examples; draw your audience's attention by using simple language, strong statements and vivid descriptions.

Ⅳ Writing

Persuasive writing

Suppose you are a member of the college newspaper with a major responsibility to write for the Wechat Official Account and manage its operation. Because of your creative reporting style that incorporates heated topics, trendy products and keen observations of students' lives with the college news, you are invited to start a new column named "No News, Just 'New's" that is sponsored by a host of Internet companies. Your new job will be to promote their new products with your own unique perspective, creative writing and detailed, vivid illustration of the product's featuring functions.

This week's sponsor is Freescale technology, and they are promoting a digital watch that meets their every claim advocated in the text. You are about to create a persuasive yet entertaining advertising article to promote this watch just for college students. You can first list at least 3 functions that will be available with the watch. (You may need to refer to the text again for helpful details.)

1. _____
2. _____
3. _____

By carefully reading the article again, you may also notice that the writer employs both facts and opinions in promotion. (Use your own words to paraphrase if possible.)

Fact	Opinion
Hospitalized patients requiring close attention can be constantly monitored using IoT technology.	Freescale provides extremely clear display of information needed continuously and it is user friendly.

Focus on how you will use examples to better illustrate its functions. You may refer to paragraph 8 to create an imitative writing:

Imagine if you are burning the midnight oil for the upcoming tests. Who haven't?

_____
_____
_____
_____

Develop the paragraph into a full article. In the article, you need to create interest for readers, transform technical details to everyday language, and make good use of exemplifications, catchy phrases and metaphors to compose an interesting piece of writing. The more details you put in illustration, the more fun people have reading it.

KEY

II Vocabulary

Task 1

a. spectrum

b. comprehensive

c. ongoing

d. preventive

e. enabling

f. well-being

h. preceding

Task 2

1. hub 2. complication 3. aggregation 4. spectrum 5. intervention
6. chronic 7. appropriate 8. intoxication 9. converge 10. incorporate

Script for Listening

This is a day I've been looking forward to for two-and-a-half years. Every once in a while, a revolutionary product comes along that changes everything. And Apple has been—well, first of all, one's very fortunate if you get to work on just one of these in your career. Apple's been very fortunate. It's been able to introduce a few of these into the world. 1984, introduced the Macintosh. It didn't just change Apple. It changed the whole computer industry. In 2001, we introduced the first iPod, and it didn't just change the way we all listen to music, it changed the entire music industry. Well, today, we're introducing three revolutionary products of this class. The first one is a widescreen iPod with touch controls. The second is a revolutionary mobile phone. And the third is a breakthrough Internet communications device. So, three things: a widescreen iPod with touch controls; a revolutionary mobile phone; and a breakthrough Internet communications device. An iPod, a phone, and an Internet communicator. An iPod, a phone … are you getting it? These are not three separate devices, this is one device, and we are calling it iPhone. Today, Apple is going to reinvent the phone, and here it is. No, actually here it is, but we're going to leave it there for now.

So, before we get into it, let me talk about a category of things. The most advanced phones are called smart phones, so they say. And they typically combine a phone plus some e-mail capability, plus they say it's the Internet. It's sort of the baby Internet, into one device, and they all have these little plastic keyboards on them. And the problem is that they're not so smart and they're not so easy to use, and so if you kind of make a Business

School 101 graph of the smart axis and the easy-to-use axis, phones, regular cell phones are right there, they're not so smart, and they're not so easy to use. But smart phones are definitely a little smarter, but they actually are harder to use. They're really complicated. Just for the basic stuff people have a hard time figuring out how to use them. Well, we don't want to do either one of these things. What we want to do is make a leapfrog product that is way smarter than any mobile device has ever been, and super-easy to use [SIC]. This is what iPhone is. OK?

So, we're going to reinvent the phone. Now, we're going to start with a revolutionary user interface. It is the result of years of research and development, and of course, it's an interplay of hardware and software. Now, why do we need a revolutionary user interface. Here's four smart phones, right? Motorola Q, the BlackBerry, Palm Treo, Nokia E62—the usual suspects. And, what's wrong with their user interfaces? Well, the problem with them is really sort of in the bottom 40 there. It's this stuff right there. They all have these keyboards that are there whether you need them or not to be there. And they all have these control buttons that are fixed in plastic and are the same for every application. Well, every application wants a slightly different user interface, a slightly optimized set of buttons, just for it. And what happens if you think of a great idea six months from now? You can't run around and add a button to these things. They're already shipped. So what do you do? It doesn't work because the buttons and the controls can't change. They can't change for each application, and they can't change down the road if you think of another great idea you want to add to this product.

Well, how do you solve this? Hmm. It turns out, we have solved it! We solved it in computers 20 years ago. We solved it with a bit-mapped screen that could display anything we want. Put any user interface up. And a pointing device. We solved it with the mouse. We solved this problem. So how are we going to take this to a mobile device? What we're going to do is get rid of all these buttons and just make a giant screen. Now, how are we going to communicate this? We don't want to carry around a mouse, right? So what are we going to do? Oh, a stylus, right? We're going to use a stylus. No. Who wants a stylus. You have to get em and put em away, and you lose em. Yuck. Nobody wants a stylus. So let's not use a stylus. We're going to use the best pointing device in the world. We're going to use a pointing device that we're all born with—born with ten of them. We're going to use our fingers. We're going to touch this with our fingers. And we have invented a new technology called multi-touch, which is phenomenal. It works like magic. You don't need a stylus. It's far more accurate than any touch display that's ever been shipped. It ignores unintended touches, it's super-smart. You can do multi-finger gestures on it. And boy, have we patented it. So we have been very lucky to have brought a few revolutionary user interfaces to the market in our time. First was the mouse. The second was the click wheel. And now, we're going to bring multi-touch to the market. And each of these revolutionary user interfaces has made possible a revolutionary product—the Mac, the iPod and now the iPhone. So, a revolutionary user interface. We're going to build on top of that with software. Now, software on mobile phones is like baby software. It's not so powerful, and today we're going to show you a software breakthrough. Software that's at least five years ahead of what's on any other phone. Now how do we do this? Well, we start with a strong foundation. iPhone runs OSX. Now, why would we want to run such a sophisticated operating system on a mobile device? Well, because it's got everything we need. It's got multi-tasking. It's got the best networking. It already knows how to power manage. We've been doing this on mobile computers for years. It's got awesome security. And the right apps. It's got everything from Cocoa and the graphics and it's got core animation built in and it's got the audio and video that OSX is famous for. It's got all the stuff we want. And it's built right in to iPhone. And that has let us create desktop class applications and networking. Not the crippled stuff that you find on most phones. This is real, desktop-class applications.

Now, you know, one of the pioneers of our industry, Alan Kay, has had a lot of great quotes throughout the years, and I ran across one of them recently that explains how we look at this, explains why we go about doing things the way we do, because we love software. And here's the quote: "People who are really serious about software should make their own hardware." Alan said that 30 years ago, and this is how we feel about it. And so we're bringing breakthrough software to a mobile device for the first time. It's five years ahead of anything on any other phone.

(https://genius.com/Steve-jobs-iphone-keynote-2007-annotated#note-1993061)

Unit 8 Flying Smarter—The Smart Airport and the Internet of Things

1 Digital technology seems to connect everything today. So it is perhaps no surprise that airports—the infrastructure that helps billions of travelers connect across the globe—are themselves becoming more digitally connected. The smart airport brings together a variety of technologies through the IoT, with the goal of strategically differentiating an airport, including via improved traveler experience, and tapping into monetary benefits through greater efficiencies and new **revenue streams**.

IoT as a technology architecture

2 IoT is a way of bringing together different enabling technologies in a specific way to do something new. But how exactly do enabling technologies come together to form an architecture? IoT's architecture is explained conceptually using the **information value loop**. However, perhaps the best way to understand it is through an example from an airport:
- The airport needs to locate its nonmotorized **ground service** equipment to maintain it more regularly. GPS tags affixed to the equipment function as sensors to create a thread of digital information about the location of a specific piece of the equipment.
- A network of radios communicates that information back to a central server.
- The server aggregates the location of one cart with the location, type, and maintenance schedule for all of the other carts.
- All this data is analyzed together to create a plan for which carts need to be retrieved today and undergo maintenance.
- With that information in hand, workers act, bringing in the right pieces of equipment for maintenance on time.

3 IoT is already in use in airports in many different ways, such as in traveler information systems, traveler traffic monitoring, baggage systems, and facilities management. Most of these uses focus on increasing efficiency. Other uses of IoT, such as enhancing security effectiveness, can improve both efficiency

---

infrastructure /ˈɪnfrəˌstrʌktʃə/ *n.* the basic systems and services that are necessary for a country or an organization to run smoothly, for example buildings, transport and water and power supplies

differentiate /ˌdɪfəˈrenʃieɪt/ *v.* to recognize or show that two things are not the same

monetary /ˈmʌnɪtəri/ *adj.* connected with money, especially all the money in a country

affix /əˈfɪks/ *v.* (formal) to stick or attach sth to sth else

retrieve /rɪˈtriːv/ *v.* (formal) to bring or get sth back, especially from a place where it should not be

(maintaining throughput with fewer machines or staff) and differentiation (shorter and faster lines for a better traveler experience). Finally, still other use cases can directly generate new revenue such as the use of **geofencing** to gather fees from **rideshares**. As a result, these three categories help us understand what IoT can achieve at an airport.

Operational efficiency

4　The majority of current airport uses of IoT focus on operational efficiency. For example, an airport representative shared that one airport has "a new online inspection system that uses a tool provided to maintenance grounds crew and connected to the Internet with GPS functionality. The purpose of this tool is to digitally connect maintenance crew inspection findings to a map of the airport grounds".

5　Another airport is planning a smart bathroom pilot test. The airport will install IoT sensors on various bathroom assets, including faucets, toilets, lighting, soap dispensers, air fresheners, toilet paper dispensers, and other equipment, in one of its busiest bathrooms. These sensors will transmit data to facilities management to alert it in real time of various shortages and breakdowns. As part of this pilot, the bathroom will have a people counter and a customer input button to capture perceptions of bathroom cleanliness, which will enable facilities management to gauge perceptions of cleanliness against actual use. This may enable greater efficiency for maintenance staff while also providing a better traveling experience.

Strategic differentiation

6　While gains in efficiency can be significant, airports that use IoT can also provide a more differentiated product or better customer experience than nonsmart airports. However, differentiation can be much broader, especially for airports where the greatest competition comes not from other airports but from other modes of travel. For example, limiting greenhouse emissions, reducing noise levels over neighboring areas, or even responsibly maintaining an airport's open space can all be differentiators that make the airport an attractive brand as well as an integral part of the community. In fact, smart products used

---

faucet /'fɔːsɪt/ *n.*
a device that controls the flow of a liquid or gas from a pipe or container

dispenser /dɪ'spensə/ *n.*
a machine or container holding money, drinks, paper towels, etc. that you can obtain quickly, for example by pulling a handle or pressing buttons

gauge /geɪdʒ/ *v.*
to measure sth accurately using a special instrument

emission /ɪ'mɪʃən/ *n.*
(formal) the production or sending out of light, heat, gas, etc.

as part of IoT can even gather information about customer preferences, providing a deeper understanding of what does and does not differentiate travel options.

7   Differentiation is fundamentally about how an airport provides distinctive value to important stakeholders. Differentiation can underpin and support the brand of an airport. IoT applications that support differentiation can come in many different varieties, aimed at traveler satisfaction or even environmental causes. For example, **Heathrow Airport** set the goal of reducing nitrogen dioxide emissions to help improve local air quality. The airport realized that a major—and avoidable—source of ground-level **nitrogen dioxide** emissions is aircraft parked at the gate that use **auxiliary power units** (APUs) instead of plugging into the power grid. As a result, Heathrow Airport deployed an IoT solution to help improve air quality. Microphones positioned around the **apron** pick up the telltale sound of APUs running. These data are **cross-referenced** against schedules and other data to determine whether an aircraft is running its APU instead of using the power grid. The airport can then share this data with airlines and remind aircraft to plug in and switch off the APU—not just saving money for the airline but also improving local air quality for all.

underpin /ˌʌndəˈpɪn/ v.
(formal) to support or form the basis of an argument, a claim, etc.

telltale /ˈtɛlteɪl/ adj.
showing that sth exists or has happened

grid /grɪd/ n.
a system of electric wires or pipes carrying gas, for sending power over a large area

New revenue

8   IoT can also create new sources of revenue. This can come from creating new products or services to attract new customers or by using IoT to sell more to existing customers. While IoT solutions aiming to generate new revenue are often the largest and most complex, they can also build upon existing solutions that generate efficiency gains to help ease implementation. Our interviews with **subject matter experts** indicate that airport stakeholders do see the value in such uses and may pursue them in the future. Use cases being investigated include variable rates for advertising and **off-airport transit** recommendations personalized to an individual traveler. However, because these revenue-generating use cases may involve the greatest number of stakeholders, they are often the hardest to pursue.

9   That said, just because use cases that create new revenue are hard to create does not mean that they have to be technologically advanced. One airport used Wi-Fi access points as sensors to

stakeholder /ˈsteɪkˌhəʊldə/ n.
a person or company that is involved in a particular organization, project, system, etc., especially because they have invested money in it

> **dwell** /dwɛl/ v.
> (formal) (literary) to live somewhere

measure the location and <u>dwell</u> time of people as they moved through the terminals. Armed with this data, the airport was able to put signs and advertising in places most likely to be seen by the right people. So while earlier, very few sales were made to landing travelers, this airport was able to place signs for the products those travelers may want to buy before departing the airport where they would be likely to see them. The result was increased sales to a previously untapped group for retailers, and through them, increasing the airport operator's **landside** revenue.

10 Regardless of the technology or application involved, IoT can have the biggest impact on airports when the technology is incorporated into the core business model. IoT might be used to facilitate a seamless door-to-door experience for air travelers: A single platform that can order and pay for rides on rail or taxi, handle travel documents, and order additional services could open up entirely new business opportunities for airport operators currently dependent on retail sales and parking receipts and greater customer interaction for airlines and other stakeholders. This **mobility-as-a-service** approach is exactly the type of IoT-based new business model that could deliver a new future to airports. Other opportunities may include <u>biometric</u> **check-in**, variable services or prices based on wait times, and individualized boarding processes.

> **biometric** /ˌbaɪəʊˈmɛtrɪk/ adj.
> using measurements of human features, such as fingers or eyes, in order to identify people

Word count: 1,216

Source: Adapted from Mariani J, Zmud J, Krimmel E, et al. Flying smarter—The smart airport and the Internet of Things [EB/OL]. (2019-07-01) [2019-08-15]. https://www2.deloitte.com/insights/us/en/industry/public-sector/iot-in-smart-airports.html.

Ⅰ Reading

Task 1

Identifying functions of the text: Annotating the text

This task focuses on the functions of different parts of the text. It will help you to understand how a text is logically constructed and how the writer develops his or her thesis.

The information in the text is organized in a particular way, where parts of the text serve different functions.

Directions: Before you read, look at the range of functions in the table and match each one with a description.

Functions	Description	Function number
1. evaluation	a. the outcomes, consequences or effects of doing something or of something happening	
2. problem(s)	b. when a new point is introduced or about to be introduced; usually occurs at the beginning or the end of a paragraph or section of text	
3. background	c. the answer to a problem, or the process of arriving at an answer	
4. summing up	d. matters that involve difficulties and need solutions	
5. cause	e. judgement or analysis of, e.g., solutions to certain problems	
6. transition	f. concluding remarks which encapsulate ideas previously presented; often found at the end of a section of text, but not always	
7. solution	g. a strong statement that presents the writer's main argument or claim to the reader	
8. result(s)	h. the reason for something happening, that has result(s) or consequence(s)	
9. thesis statement	i. information that sets the scene and/or looks at an existing situation; usually at the beginning of a section or text, but not always	3

Directions: Complete the annotation of the text with the given example. This will show how you can annotate the text by highlighting relevant parts according to their functions.

1. Re-read the text, paying close attention to the two annotations already made.

2. Locate, highlight and annotate other parts of the text that demonstrate the remaining functions from the table.

3. Compare your annotations with another student's. Then compare them with a model your teacher will provide.

Annotation	Text
Background	Digital technology seems to connect everything today. So it is perhaps no surprise that airports—the infrastructure that helps billions of travelers connect across the globe—are themselves becoming more digitally connected. The smart airport brings together a variety of technologies through the IoT, with the goal of strategically differentiating an airport, including via improved traveler experience, and tapping into monetary benefits through greater efficiencies and new revenue streams.
Further information	
Definition	**IoT as a technology architecture** IoT is a way of bringing together different enabling technologies in a specific way to do something new. But how exactly do enabling technologies come together to form an architecture? IoT's architecture is explained conceptually using the information value loop. However, perhaps the best way to understand it is through an example from an airport:
Exemplification	

Task 2

Identifying the writer's purpose: Annotating the text

It is important to identify what the writer's purpose is and his/her attitude to the topic. This helps you decide about the content and the extent to which it meets your needs. Go through the rest of the text and make brief annotation in the margin. Use a pencil in case you decide to make changes.

Directions: Identify which sections of the text have the following functions (not all of these functions might be used in this text).

- background information, e.g., giving detail of the overall situation
- clarification, e.g., making a statement more comprehensible
- concession, e.g., making a statement less strong
- explanation
- exemplification, i.e., examples
- general problems/implications
- evaluation
- conclusion
- viewpoints, i.e., writer's recommendations or suggestions
- solution
- explication, i.e., further information to develop a point

Ⅱ Vocabulary

Key words and expressions

infrastructure	differentiate	monetary
affix	retrieve	faucet
dispenser	gauge	emission
underpin	telltale	grid
stakeholder	dwell	biometric

Chapter 2 The Applications of the IoT

Terminology

revenue stream	n.	收入来源
information value loop	n.	信息值循环
ground service	n.	地勤
geofencing	n.	地理围栏
rideshare	n.	共乘
Heathrow Airport	n.	希思罗机场
nitrogen dioxide	n.	二氧化氮
auxiliary power units (APUs)	n.	辅助动力单元
apron	n.	停机坪
cross-referenced	adj.	前后参照的
subject matter experts (SMEs)	n.	领域专家
off-airport transit	n.	机场外交通
landside	n.	机场公共场所
mobility-as-a-service	n.	出行服务
check-in	n.	登机手续办理

Task 1

Directions: Fill in the blanks with prepositions that complete best collocations.

1. The Campbell Soup Company says it will try to tap _____ Japan's rice market.

2. In China, there is no other city like Shanghai which brings _____ a number of colorful lifestyles.

3. For sedans, posters can be affixed _____ the car body and the rear window.

4. Having fully searched the crime scene, the police had the situation well _____ hand.

5. I have three part-time jobs, which bring _____ about $24,000 a year.

6. My wife has a lot of friends who are plugged _____ various performance arts around the city.

7. Whenever he starts in on economics, I switch _____ automatically.

8. The prince had incorporated himself _____ the main governing body.

9. The company is so massive that there are jobs opening _____ all the time.

Task 2

Directions: Look at the text to find the text-referring words in the table. Note down the idea or word(s) that each one refers to.

Para(Line)	Text-referring word(s)	Refers to
3(3)	these	different ways IoT is used in airports, such as in traveler information systems, traveler traffic monitoring, etc.

Continued

Para(Line)	Text-referring word(s)	Refers to
3(10)	these three	
5(11)	this	
7(13)	these	
8(5)	they	
9(1)	That	
9(10)	they	
9(10)	them	
9(12)	them	

Ⅲ Listening and Speaking

Task 1

Directions: Scan the QR code on the margin and you will listen to the CNN reports on how to protect yourself from hackers, and fill in the blanks when you listen for the second time.

Most of us are walking a tightrope when it comes to protecting ourselves from the hackers. Living on the edge, thinking, oh, I won't fall. That's why it's so important to protect yourself.

Report

Just like you do a safety check before you (1)_____ a big stunt, you need to do a (2)_____ on your computer to make sure you are running the latest version of your operating system. And you always have to do a (3)_____ of your most important files just in case. But if all that security lets you down, encrypt your files with FileVault or BitLocker. And to be on (4)_____, always run antivirus software. I'm going from log to log, but you probably feel like you're going from login to login, password to password. There are so many opportunities for the hackers to (5)_____ those. So you should use (6)_____ sign-in. That way, every time you have to enter your password, you also get a code sent to your phone. So even if the hackers steal your password, they won't get their hands on your smartphone.

And while we're on the logs, popular messaging apps like WhatsApp and iMessage—they keep logs of your text messages up in the cloud. And you know that gets hacked every so often. But apps like Signal—they don't save anything in the cloud, making them a much safer place for your text messages to (7)_____.

To make sure that surfing the web is all (8)_____, check that websites you visit have HTTPS. That "S" means that they have (9)_____ security. But if you're (10)_____ logging on to public Wi-Fi, VPN apps like Onavo will make your connection private. And for the softest surfing, the web browser Tor will make it so that somebody can't know who you are, where you are, or what sites you're surfing.

Living on the edge doesn't have to be so dangerous if you use the right protection.

Task 2

Directions: Survey in your class on the same topic, and also collect other relevant information such as when, where, how and why people are most likely to be hacked, and inquire classmates their further opinions on being hacked. You can either arrange interviews, send out questionnaires or reflect on personal experiences. Organize all the materials collected from your survey, and prepare for a 5 minute presentation to share what you have found. It is recommended that you display the complete process of your surveying and analyzing and work out answers in a limited scope with a unique perspective.

Ⅳ Writing

Task Writing a report

Each semester, you will probably be asked by at least one instructor to read a book or an article and write a paper recording your response to the material. In the reports or reaction papers, your instructor will most likely expect you to do two things: summarize the material and detail your reaction to it.

1. Summary writing

Read the article for all you can understand the first time through. Pay special attention to the content relevant to your predictions. Also, look for definitions, examples and enumerations (lists of items), which often indicate key ideas. You can also identify important points by turning any headings into questions to find the answers to the questions.

Identify the author and title of the text, and include in parentheses the publisher and publication date.

Write an informative summary of the material. Condense the content of the work by highlighting its main points and key supporting points. Do not discuss in great detail any single aspect of the work. Also, keep the summary objective and factual.

2. Writing your reaction to the work

Focus on any or all of the questions below. (Check with your instructor to see whether you should emphasize specific points.)

a. How is the article related to ideas and concerns discussed in the course? For example, what points made in the course textbook, class discussions, or lectures are treated more fully in this article?

b. How is the article related to problems in our present-day world?

c. How is the work related to your life, experience, feelings, and ideas? For instance, what emotional reactions did you have? Did it increase your understanding of an issue or change your perspective?

Evaluate the merit of the work: the importance of its points; its accuracy, completeness, and organization; and so on. You should also indicate here whether you would recommend the article to others, and why.

KEY

I Reading

Task 1

a. 8 b. 6 c. 7 d. 2 e. 1 f. 4 g. 9 h. 5

II Vocabulary

Task 1

1. into 2. together 3. to 4. in 5. in 6. into 7. off 8. into 9. up

Task 2

Para(Line)	Text-referring word(s)	Refers to
3(3)	these	different ways IoT is used in airports, such as in traveler information systems, traveler traffic monitoring, etc.
3(10)	these three	three ways IoT is used in airports with different focuses on increasing efficiency, enhancing both efficiency and differentiation, and directly generating new revenue
5(11)	this	smart bathroom pilot test that includes a people counter and a customer input button to gauge perceptions of cleanliness against actual use
7(13)	these	sound of APUs running
8(5)	they	IoT solutions aiming to generate new revenue
9(1)	That	revenue-generating use cases are often the hardest to pursue
9(10)	they	landing travelers
9(10)	them	products landing travelers may want to buy
9(12)	them	increasing sales to a previously untapped group for retailers

III Listening and Speaking

Task 1

(1) perform (2) security check (3) backup (4) solid ground (5) get their hands on (6) two-step (7) land (8) smooth sailing (9) higher-level (10) feeling shaky about

Script for Listening

Most of us are walking a tightrope when it comes to protecting ourselves from the hackers. Living on the edge, thinking, oh, I won't fall. That's why it's so important to protect yourself.

Just like you do a safety check before you perform a big stunt, you need to do a security check on your computer to make sure you are running the latest version of your operating system. And you always have to do a backup of your most important files just in case. But if all that security lets you down, encrypt your files with FileVault or BitLocker. And to be on solid ground, always run antivirus software. I'm going from log to log, but you probably feel like you're going from login to login, password to password. There are so many opportunities for the hackers to get their hands on those. So you should use two-step sign-in. That way, every time you have to enter your password, you also get a code sent to your phone. So even if the hackers steal your password, they won't get their hands on your smartphone.

And while we're on the logs, popular messaging apps like WhatsApp and iMessage—they keep logs of your text messages up in the cloud. And you know that gets hacked every so often. But apps like Signal—they don't save anything in the cloud, making them a much safer place for your text messages to land.

To make sure that surfing the web is all smooth sailing, check that websites you visit have HTTPS. That "S" means that they have higher-level security. But if you're feeling shaky about logging on to public Wi-Fi, VPN apps like Onavo will make your connection private. And for the softest surfing, the web browser Tor will make it so that somebody can't know who you are, where you are, or what sites you're surfing.

Living on the edge doesn't have to be so dangerous if you use the right protection.

Unit 9 Industrial Internet of Things

What is the Industrial Internet of Things (IIoT)?

1 The Industrial Internet of Things or IIoT initially mainly referred to an industrial framework whereby a large number of devices or machines are connected and synchronized through the use of software tools and third platform technologies in an M2M and IoT context, later an **Industry 4.0** or Industrial Internet context.

2 Today it is mainly used in the scope of IoT applications outside of the consumer space and is about applications and use cases across several sectors, to distinguish between consumer IoT applications and business/industry applications.

3 IIoT is defined as "machines, computers and people enabling intelligent industrial operations using advanced data analytics for transformational business outcomes".

M2M communication and the Industrial Internet of Things

4 In the pure M2M and Industry 4.0 context, the advantage of the frameworks and systems that IIoT refers to, is that they can operate semi-independently or with very minimal human intervention.

5 Such systems will increasingly be able to intelligently respond and even change their course of action based on the information received through the **feedback loops** established within the framework.

6 As mentioned, a keyword here is M2M communication. The idea behind M2M communication is to reduce human interventions as much as possible so that the highest level of **automation** could be achieved.

7 If we look at the concept of the Internet of Everything, this M2M dimension of IIoT happens within the sphere of the things as you can see in the original depiction of the Internet of Everything by **Cisco**.

8 IIoT in this sense can be considered a movement towards "smart machines" whereby the accuracy levels of the operations

---

whereby /wɛəˈbaɪ/ *adj.*
by which
synchronize /ˈsɪŋkrənaɪz/ *v.*
operate simultaneously

analytics /ˌænəˈlɪtɪks/ *n.*
the discovery and communication of meaningful patterns in data
transformational /ˌtrænsfəˈmeɪʃənəl/ *adj.* changed
outcome /ˈaʊtˌkʌm/ *n.*
a phenomenon that follows and is caused by some previous phenomenon

sphere /sfɪə/ *n.*
a particular environment or walk of life
depiction /dɪˈpɪkʃən/ *n.*
a graphic or vivid verbal description
accuracy /ˈækjʊrəsi/ *n.*
the quality of being near to the true value

involved in the respective systems are heightened to a level that cannot be achieved through human interventions.

Benefits of IIoT in manufacturing and beyond

9 One of the greatest benefits of IIoT has to be seen in the reduction of human errors and manual labor, the increase in overall efficiency and the reduction of costs, both in terms of time and money. We also cannot forget the possible underpinnings of IIoT in quality control and maintenance.

10 IIoT is part of the IoT. IoT is data-rich: large amounts of data get collected, aggregated and shared in a meaningful way. Here again the goal is to increase the automation level at domestic and commercial levels. In the IIoT, data is crucial as well and this causes a change in the human tasks in an Industry 4.0 context whereby automation leads to a decrease of specific types of work but at the same time requires new skillsets. The goal of IIoT is also not to fully replace human work; its goal is to enhance and optimize it by, for example, creating new revenue streams and business models with a big role for data (analysis).

11 The **intelligent communication loop** setup between machines enables timely attention to maintenance issues. The safety level of the operations is also boosted by alleviating the risk factors.

12 IIoT takes the benefits of the IoT in general to a higher level and also to the industries with high-stakes where human error could result in massive risks. The precision level that can be achieved through IIoT is one of the greatest advantages, which makes this discipline one of the most welcome gifts of IoT.

13 Times are not far whereby entire manufacturing plant operations and processes could be made to operate almost independently. Moreover, IIoT is used for many use cases which help us reduce the exposure of human workforce, which will always matter, to scenarios with high industrial hazards.

14 In the coming years, IIoT is likely to force more unified device protocols and architectures that will allow machines to communicate seamlessly and thereby enhance interoperability.

15 To summarize, here are some of the key benefits of IIoT in an industry context:

- Improved and intelligent connectivity between devices or machines;
- Increased efficiency;

---

heighten /ˈhaɪtən/ v.
make more extreme; raise in quantity, degree, or intensity

manual /ˈmænjʊəl/ adj.
doing or requiring physical work
in terms of in the field of
underpinning /ˌʌndəˈpɪnɪŋ/ n.
a solid foundation land below ground level to support or strengthen a building
aggregate /ˈæɡrɪɡɪt/ v.
gather in a mass, sum, or whole
domestic /dəˈmɛstɪk/ adj.
connected with the home or family
skillset /ˈskɪlˌsɛt/ n.
a particular category of knowledge, abilities, and experience necessary to optimize a job
stream /striːm/ n.
a steady flow
alleviate /əˈliːvɪˌeɪt/ v.
to make sth less severe
stakes /steɪks/ n.
the money risked on a gamble
precision /prɪˈsɪʒn/ n.
the quality of being exact, accurate and careful
exposure /ɪkˈspəʊʒə/ n.
the state of being in a place or situation where there is no protection from sth harmful or unpleasant
workforce /ˈwɜːkˌfɔːs/ n.
the total number of people in a country or region who are physically able to do a job and are available for work
scenario /sɪˈnɑːrɪˌəʊ/ n.
a description of how things might happen in the future
hazard /ˈhæzərd/ n.
a thing that can be dangerous or cause damage

- Cost savings;
- Time savings;
- Enhanced industrial safety.

16　It is certainly a worthwhile space to watch as the developments here bring about the biggest industry revolution of the Internet era but also because most IoT deployments today happen in IIoT and, specifically in manufacturing and factory environments.

IIoT protocols

17　One of the issues encountered in the transition to IIoT is the fact that different **edge-of-network** devices have historically used different protocols for sending and receiving data. While there are a number of different communication protocols currently in use, such as **OPC-UA**, the Message Queueing Telemetry Transport (MQTT) transfer protocol is quickly emerging as the standard for IIoT, due to its lightweight overhead, publish/subscribe model, and bidirectional capabilities.

Challenges of IIoT

18　Interoperability and security are probably the two biggest challenges surrounding the implementation of IIoT. As technology writer Margaret Rouse observes, "A major concern surrounding IIoT is interoperability between devices and machines that use different protocols and have different architectures." Now there are already excellent solution for this. Ignition is one since it is cross-platform and built on open-source, IT-standard technologies.

19　Companies need to know that their data is secure. The proliferation of sensors and other smart, connected devices has resulted in a parallel explosion in security vulnerabilities. This is another factor in the rise of MQTT since it is a very secure IIoT protocol.

The Future of IIoT

20　IIoT is widely considered to be one of the primary trends affecting industrial businesses today and in the future. Industries are pushing to modernize systems and equipment to meet new

---

deployment /dɪˈplɔɪmənt/ n.
the organization and positioning of troops, resources, or equipment so that they are ready for quick action

transition /trænˈzɪʃn/ n.
the process or a period of changing from one state or condition to another

lightweight /ˈlaɪtweɪt/ adj.
weighing less than most other things of the same type

subscribe /səbˈskraɪb/ v.
to pay an amount of money regularly in order to receive or use sth

proliferation /prəˌlɪfəˈreɪʃn/ n.
the sudden increase in the number or amount of sth

parallel /ˈpærəlel/ adj.
(events or situations) happen at the same time as one another

vulnerability /ˌvʌlnərəˈbɪləti/ n.
the state of being easily harmed or affected by something bad

modernize /ˈmɒdənaɪz/ v.
to make a system, methods, etc. more modern and more suitable for use at the present time

regulations, to keep up with increasing market speed and volatility, and to deal with disruptive technologies. Businesses that have embraced IIoT have seen significant improvements to safety, efficiency, and profitability, and it is expected that this trend will continue as IIoT technologies are more widely adopted.

Word count: 953

Source: What is IIoT? The Industrial Internet of Things [EB/OL]. (2018-07-13)［2019-02-11］. https://inductiveautomation. com/resources/article/what-is-iiot.

IIoT—the Industrial Internet of Things (IIoT) explained [EB/OL]. ［2019-02-11］. https://www. i-scoop. eu/internet-of-things-guide/industrial-internet-things-iiot-saving-costs-innovation/industrial-internet-things-iiot/.

volatility /ˌvɒləˈtɪləti/ *n.*
the state of being likely to change suddenly and unexpectedly

disruptive /dɪsˈrʌptɪv/ *adj.*
causing problems, noise, etc. so that sth cannot continue normally

profitability /ˌprɒfɪtəˈbɪləti/ *n.*
the quality of being able to make a profit

Ⅰ Reading

Catching signal words

People say "words are more powerful than swords". All words are powerful, but some words have superpowers. They are called "signal words". Signal words are specific words that you can use to transition between the different ideas in your paper clearly and organically. The ability to identify and understand the meaning of these words is crucial in any kind of reading—for fun, and for academic purpose as well.

The following chart lists some commonly used categories of signal words.

Function	Examples
Addition	also, moreover, furthermore, and then, in addition, together with, further, likewise, equally important, along with, plus, besides, too, over and above, as well as, another, next
Comparison	similarly, in like manner, just as, identically, likewise, similar to
Contrast	but, yet, however, after all, nevertheless, in contrast, though, on the contrary, otherwise, whereas, although, despite, on the other hand, different from, even though, instead, unlike
Cause and Effect	because, for, since, if, on account of, therefore, so, accordingly, then, for this reason, thus, consequently, hence, thus
Exemplification	for example, for instance, to illustrate, specifically, such as
Summary	to sum up, in short, as has been noted, in brief, as I have said, in other words, on the whole, in fact, that is, indeed, finally, in summary, to be sure, in any event, for these reasons, thus, above all
Time	first, second, third, next, last, since, formerly, previously, at length, at last, soon, meanwhile, afterward, often, now, after, finally, until, during, following, when, while, then

Of course there are much more categories that can be listed here, among which are introduction, classification, concession, repetition, place, etc. You may accumulate them through your process of reading.

Task 1

Directions: Please read paragraphs 9 and 10 again, find as many signal words as you can, and decide which category each of them belongs to.

word	category	word	category
_____	_____	_____	_____
_____	_____	_____	_____
_____	_____	_____	_____
_____	_____	_____	_____
_____	_____	_____	_____

Reading and understanding the text

Task 2 Multiple choice

Directions: Please answer the following questions by choosing the right answers from the four choices marked A, B, C and D.

1. Which of the following is NOT correct about the definition of IIoT? (   )

   A. In IIoT, a large number of devices or machines are connected and synchronized.

   B. It is used in the scope of IoT applications outside of the consumer space.

   C. It is machines, computers and people enabling intelligent industrial operations.

   D. It is about applications synchronizing consumer and industry IoT applications.

2. What is the idea behind M2M communication? (   )

   A. To connect as many machines as possible

   B. To reduce human interventions as much as possible

   C. To reduce cost and time as much as possible

   D. To achieve complete automation

3. Which is the key benefit of IIoT in an industry context? (   )

   A. Improved and intelligent connectivity between devices or machines

   B. Increased efficiency and enhanced industrial safety

   C. Cost savings and time savings

   D. All the above

4. Which of the following is CORRECT about the outlook of IIoT? (   )

   A. Interoperability is a big challenge which will stop IIoT from developing.

   B. There is nothing serious about security in future IIoT development.

   C. Businesses and companies who adopt IIoT would earn more money.

   D. IIoT would push to modernize equipment, which may slow down the market increase.

5. Which is the most important standard for IIoT nowadays? (   )
A. MQTT   B. OPC-UA   C. M2M   D. Industry 4.0

II Vocabulary

Key words and expressions

whereby	synchronize	analytics	transformational
outcome	sphere	depiction	accuracy
heighten	manual	in terms of	underpinning
aggregate	domestic	skillset	stream
alleviate	stakes	precision	exposure
workforce	scenario	hazard	deployment
transition	lightweight	subscribe	proliferation
parallel	vulnerability	modernize	volatility
disruptive	profitability		

Terminology

Industry 4.0	n.	工业 4.0
feedback loop	n.	反馈回路
automation	n.	自动化
Cisco	n.	思科公司
intelligent communication loop	n.	智能通信回路
edge-of-network	n.	边缘网络
OPC-UA(object linking and embedding for process control-unified architecture)	n.	用于过程控制的对象链接与嵌入-统一架构
overhead	n.	（企业或设备的）日常开销

Task 1 Word bank

Directions: Please fill in the blanks with the words listed below. You may need to change the form when necessary.

whereby	in terms of	precision	subscribe
alleviate	aggregate	exposure	vulnerability
depiction	accuracy	scenario	disruptive

1. _____ unemployment, the troubles of the Detroit auto industry have hit Michigan in the Midwest especially hard.

2. The cartoon vividly _____ the relationship between wealth and happiness.

3. Taking long-term courses of certain medicines can make patients _____ to infection.

4. This machinery is so _____ that a human hair couldn't even pass between the heads and spinning platters, yet it all works in terrific speeds.

5. Alcohol can produce violent, _____ behavior.

6. Knowing where, when, and how to move will require rigorous market analysis with _____ data.

7. Tranquilizers help _____ the distressing symptoms of anxiety.

8. The conflict degenerating into civil war is everybody's nightmare _____.

9. Usenet is a collection of discussion groups, known as newsgroups, to which anybody can _____.

10. The scores were _____ with the first round totals to decide the winner.

Task 2　Thesaurus

A thesaurus is a reference book in which words with similar meanings are grouped together (containing synonyms and sometimes antonyms). The main purpose of such reference works is for users to find the exact word to get their idea expressed most clearly and accurately. Unlike a dictionary, a thesaurus entry does not give the definition of words.

Please write down as many words or phrases which have similar (not exactly the same) meaning with the following ones as possible.

1. **optimize** _____
2. **disruptive** _____
3. **sphere** _____
4. **stream** _____
5. **heighten** _____
6. **hazard** _____
7. **expose** _____
8. **transform** _____

Ⅲ　Listening and Speaking

Directions: Scan the QR code on the margin and you will watch a video clip about IIoT.

Listening preparation—vocabulary

Video Clip

Please study the following new words which are going to appear in the listening material.

broad strokes		(idiomatic) vague or non-specific terms; major features or key points
instrumentation	*n.*	instruments for a specific purpose
activate	*v.*	to set in motion; make active or more active
nutrient	*n.*	a source of nourishment, especially a nourishing ingredient in a food

moisture	n.	water or other liquid diffused as vapor or condensed on or in objects
hub	n.	the central portion of a wheel, propeller, fan, etc., through which the axle passes
thermostat	n.	a device that maintains a system at a constant temperature
endpoint	n.	the entity at one end of a connection
definite	adj.	easily or clearly seen or understood; obvious
stiff	adj.	difficult, rigid
leverage	v.	to make strategic use of sth to accomplish some purpose; exploit
gadget	n.	a small tool or device that does sth useful
DDoS		distributed denial-of-service(attack)
vector	n.	a quantity that has both size and direction

Task 1 Spot dictation

Directions: Please listen to the part (00:18~01:08) of the video, and fill in the blanks below. You may listen to the video several times to get the correct words.

One IIoT system could be as simple as a (1) _____ rat trap that texts home to say that it's been activated, while another might be as complicated as a fully (2) _____ mass production line that tracks maintenance, productivity and even ordering and shipping information across a huge, multi-layered network.

Other examples can (3) _____ from things like soil sensors in a connected agriculture setup (4) _____ nutrient and moisture data to a central hub for mapping and planning, to a production line full of robots sending detailed information about automobile production and warning an operator when important components are showing (5) _____ of wear and tear. Automotive (6) _____ management, manufacturing and the energy sector are probably the biggest users of IIoT technology at this point, but retail and healthcare are seeing (7) _____ growth—retailers like the ability to track customers in their stores and (8) _____ location-relevant content (read: advertisements) in a (9) _____ way, while healthcare providers can make sure imaging and data from items like CT scanners and digital charts can (10) _____ _____ freely.

Task 2 Answer the questions

1. Can you guess the meaning of "wear and tear"?
_____
_____

2. What does the speaker mean by saying that "the scale of it (IIoT) is much different than a simple system that lets you mess with your thermostat on your phone"?
_____
_____

3. What is one of the definite issues around IIoT mentioned in the video that need to be solved?
_____
_____

4. Besides DDoS attack, what could be another security threat?

_____
_____

Task 3   Hacker on track

Directions: At the end of the video, it mentioned potential security problems of IIoT. Now please work in pairs; one of you should be the hacker and say what you can do with the less secured items in an IIoT, and then the other should think of measures to prevent or remedy the threat mentioned. Think about as many possible situations as you could, and report to the class together.

IV Writing

Making a research paper outline

The purpose of an outline is to help you think through your topic carefully and organize it logically before you start writing. A good outline is the most important step in writing a good paper. It helps you to make sure that all the points are covered and flow logically from one to the other.

In making an outline, you should first include all the basic structures of your paper, which are your title, abstract, introduction, body part (which should include your materials and methods, results and discussion), conclusion, reference and possibly acknowledgement.

An outline might be formal or informal. An informal outline (working outline) is a tool helping an author put down and organize their ideas. It is subject to revision, addition and canceling, without paying much attention to form. It helps an author to make their key points clear for him/her and arrange them.

Sometimes you are asked to submit formal outlines with your research papers.

In a formal outline, there are several common styles. The most common type is an outline with letters or (and) numbers. In which you may:

- use Roman letters to represent the heading with the main topics;
- use uppercase letters for subtopics;
- use Arabic numbers to represent details (part of a subtopic);
- use lowercase letters for points within subparts.

In academic writing, it's popular to following styles like APA and MLA. Each of these styles has its peculiar features and rules to follow. You should be familiar with it before writing. And if making an outline is a part of your assignment, follow the instructions you were given.

(https://www.wiseessays.com/blog/research-paper-outline

https://www.aresearchguide.com/1steps.html)

Task 1   Making an outline

Directions: Here are three topics. Learn about the information of each topic and choose one to make an outline for your composition.

Topic 1: An introduction to IoT-supported smart home

Topic 2: How does IoT change retail?

Topic 3: My plan for IoT smart school

Task 2 Writing

Directions: Finish your composition on the basis of the outline you have written in Task 1. (100~200 words)

KEY

I Reading

Task 1

word	category	word	category
one of	Example	cause	Cause and Effect
both	Addition	lead to	Cause and Effect
and	Addition	but	Contrast
here again	Repetition	also	Addition
as well	Addition	for example	Exemplification

Task 2

1. D 2. B 3. D 4. C 5. A

Ⅱ Vocabulary

Task 1

1. In terms of   2. depicts   3. vulnerable   4. precise   5. disruptive
6. accurate   7. alleviate   8. scenario   9. subscribe   10. aggregated

Task 2

1. **optimize**   amend,increase,develop,improve,raise
2. **disruptive**   disturbing,troublesome,unruly,upsetting,disorderly
3. **sphere**   circle,field,realm,scope
4. **stream**   current,flood,flow,rush,spate
5. **heighten**   add to,boost,enhance,magnify,strengthen,amplify
6. **hazard**   peril,risk,threat,endangerment,double trouble,hot potato
7. **expose**   uncover,unmask,display,disclose,bring to light
8. **transform**   alter,convert,mold,switch,mutate,reconstruct,remodel

Ⅲ Listening and Speaking

Task 1

(1) connected   (2) automated   (3) range   (4) relaying   (5) signs
(6) fleet   (7) rapid   (8) deliver   (9) targeted   (10) move around

Task 2

1. "Wear and tear" means damage, depreciation, or loss resulting from ordinary use.

2. He means that the scale of IIoT is much larger than the IoT engaged in everyday life, such as smart home system.

3. It's difficult to make the best use of all the data that an IIoT setup can generate; and it's also difficult to meet the requirements for device service life and reliability.

4. Those poorly secured devices can allow access to valuable data already on your network.

Script for Listening

IIoT is short for Industrial Internet of Things. In the broad strokes, it's the application of instrumentation and connected sensors and other devices to machinery and vehicles in the transport, energy and industrial sectors. What that means in practice varies widely.

One IIoT system could be as simple as a connected rat trap that texts home to say that it's been activated, while another might be as complicated as a fully automated mass production line that tracks maintenance, productivity and even ordering and shipping information across a huge, multi-layered network. Other examples can range from things like soil sensors in a connected agriculture setup relaying nutrient and moisture data to a central hub for mapping and planning, to a production line full of robots sending detailed information about automobile production and warning an operator when important components are showing signs of wear and tear. Automotive fleet management, manufacturing and the energy sector are probably the biggest users of IIoT technology at this point, but retail and healthcare are seeing rapid growth—retailers like the ability to track customers in their stores and deliver location-relevant content (read: advertisements) in a targeted way, while healthcare providers can make sure imaging and data from items like CT scanners and digital charts can move around freely.

Technologically, IIoT works on similar principles to any other piece of IoT tech—automated instrumentation and reporting being applied to stuff that didn't have those capabilities before. That said, the scale of it is much different than a simple system that lets you mess with your thermostat on your phone—hundreds, perhaps thousands or even tens and hundreds of thousands of individual endpoints can be present in an IIoT deployment. There are still definite issues around IIoT that need to be solved—making the best use of all the data that an IIoT setup can generate isn't simple, and requirements for device service life and reliability are frequently stiff.

And just like consumer IoT, IIoT has a lot of security issues. By now, everyone's heard of the Mirai botnet, which leveraged poorly secured security cameras and other gadgets into a huge DDoS weapon. Even beyond that, there's also the issue that those devices can allow access to valuable data already on your network—yet another attack vector.

Unit 10   The IoT Data Opportunity for Logistics Companies Is here

logistics /ləˈdʒɪstɪks/ *n.*
the commercial activity of transporting goods to customers
intriguing /ɪnˈtriːɡɪŋ/ *adj.*
very interesting because of being unusual or mysterious
pursue /pəˈsjuː/ *v.*
to follow someone or something, usually to try to catch him, her, or it
transform /trænsˈfɔːm/ *v.*
to change completely
commission /kəˈmɪʃən/ *v.*
to formally choose someone to do a special piece of work
respondent /rɪˈspɒndənt/ *n.*
a person who answers a request for information
roadblock /ˈrəʊdˌblɒk/ *n.*
a temporary structure put across a road to stop traffic
vendor /ˈvɛndə/ *n.*
someone who is selling something
myriad /ˈmɪriəd/ *n.*
a very large number of something
at hand near in time or position
enormous /ɪˈnɔːməs/ *adj.*
extremely large
daunt /dɔːnt/ *v.*
to make someone feel slightly frightened or worried
scalable /ˈskeɪləbəl/ *adj.*
used to describe a business or system that is able to grow or to be made larger
navigate /ˈnævɪˌɡeɪt/ *v.*
to direct the way that a ship, aircraft, etc. will travel
dense /dɛns/ *adj.*
having parts that are close together so that it is difficult to go or see through
stitch /stɪtʃ/ *v.*
to sew two things together, or to repair something by sewing
relentless /rɪˈlɛntlɪs/ *adj.*
continuing in a severe or extreme way
conceive /kənˈsiːv/ *v.*
to invent a plan or an idea
deliver /dɪˈlɪvə/ *v.*
to give or produce a speech or result

1   The Internet of Things (IoT) has moved rapidly from intriguing concept to key strategy objective that companies know they must pursue to transform their business and customer experience. In fact, a Forrester survey commissioned by **Arm** in 2018 has found more than 60 percent of respondents believe their IoT investments have or will fundamentally change their business models. More than half already are seeing improvements in operational efficiency.

2   But many companies not that far down the **deployment path** run into a roadblock, namely complexity: The complexity of what's required for a useful solution; the complexity of the sheer number of vendors and technologies to consider; the complexity of the interoperability of data types and myriad of **connectivity** choices and how to connect everything securely into their legacy **ERP systems.**

IoT opportunity is at hand

3   In my conversations with customers in the logistics world including retailers, transportation, and **3PLs** (third-party logistics), they know there's enormous value in embracing IoT and leveraging the vast amounts of **untapped data** that await them to transform their business. But they are daunted by where to begin. They know they want solutions that are flexible and scalable, secure and efficient, but they have little patience for navigating a dense forest of options to stitch together an IoT solution for themselves. In short, they're looking for packaged solutions and platforms that can speed their **data-insight**. And they want to understand the **ROI** (return on investment) for their business.

4   Arm has worked relentlessly for years to conceive and deliver

hardware **IP solutions** to support companies' IoT ambitions. More recently—seeing a need for solutions that span hardware and software—Arm launched the Pelion IoT platform. Pelion IoT platform enables organizations to connect and manage their IoT devices and data, removing much of the friction and confusion from the process of building an IoT solution. In short, it is a platform customers yearn for.

5   For developers and business leaders, Pelion IoT platform accelerates the **time to value** of IoT deployments by helping you easily connect trusted IoT devices on global networks, seamlessly administer them, and extract **real-time data** from them to drive competitive advantage.

6   To date, Arm's IoT services manage more than 30 **petabytes** of customer data, handle more than 300,000 queries a day and ingest more than 2 million records. The more than 900 customers that use the services have more than 140 Arm ecosystem partners and 350,000 developers to help them along their IoT journey. This has been recognized broadly: Arm Treasure Data was recently added to the **Gartner Magic Quadrant** analysis (published Feb. 13, 2018 by Adam Ronthal, Roxane Edjlali, and Rick Greenwald) of data-management and analytics companies.

Logistics firms embracing IoT

7   In the logistics space, customers want real-time visibility on shipments with fewer **barcode scans**. They want to move from just tracking things from point A to point B, to capturing and tracking more data, seamlessly managing all the different data types and connecting this **treasure trove** of data into their existing ERP systems.

8   Retailers are looking to IoT to have better shopper data, both in the physical and digital stores. Merchandising and marketing functions in retail are evolving to become more effective in the **omnichannel** world: A **McKinsey** survey found that two-thirds of their time was spent gathering data, managing expectations and sitting in meetings; only one third of their time was invested in strategy, analytics and insights.

9   Leading-edge customers, such as **Shiseido**, the world's third-largest cosmetics company, use Arm's data platform to ingest, integrate, store, prepare and then discover vast amounts of data, including **loyalty membership** data, second-party data such as

---

span /spæn/ v.
to cover or extend over an area or time period

launch /lɔːntʃ/ v.
to introduce something new such as a product

enable /ɪnˈeɪbəl/ v.
to make someone able to do something

friction /ˈfrɪkʃən/ n.
disagreement or unfriendliness caused by disagreements

confusion /kənˈfjuːʒən/ n.
a situation in which people do not understand what is happening

yearn /jɜːn/ v.
to wish very strongly

accelerate /əkˈsɛləˌreɪt/ v.
to happen or make something happen sooner or faster

query /ˈkwɪərɪ/ n.
a question, often expressing doubt about something or looking for an answer from an authority

ingest /ɪnˈdʒɛst/ v.
absorb, take in, consume

visibility /ˌvɪzɪˈbɪlɪtɪ/ n.
the degree to which something is seen by the public

shipment /ˈʃɪpmənt/ n.
a large amount of goods sent together to a place, or the act of sending them

retailer /ˈriːteɪlə/ n.
a person, shop, or business that sells goods to the public

merchandise /ˈmɜːtʃənˌdaɪz/ v.
to encourage the sale of goods by advertising them or by making certain that they are noticed

leading-edge /ˌliːdɪŋˈɛdʒ/ adj.
in or at the most advanced position in an area of activity

cosmetics /kɒzˈmɛtɪks/ n.
substances put on the face or body that are intended to improve its appearance or quality

media website viewing data, and third-party **public data-management platform** (**DMP**) information.

10　This embrace has improved Shiseido's **CRM** (customer relationship management) engagement and ad performance, given its teams improved data analysis and visualization and delivered a 20 percent increase in revenue from loyalty members. This led to an 11 percent increase in overall revenue.

11　Logistics-solutions provider **Zebra Technologies**' real-time **load-monitoring** solution SmartPack Trailer installed at warehouse doors helps optimize loading and ensure the right package ends up in the right truck. Additionally, Zebra is using the Pelion IoT Platform to manage **ruggedized** mobile Android computers used by large retailers around the globe.

12　These and countless other companies are moving quickly to exploit the vast benefits that the connected IoT will bring to help them unlock data to transform their businesses.

13　A new report from SJ Consulting lays out not only the challenges that this transformational technology brings but the pathways to exploiting IoT to improvement businesses. In it, you'll find:

- analysis of how the retail, technology, transportation and logistics industries are merging in response to Amazon;
- details on new competitive threats industry incumbents are facing;
- insight on the strategic partnerships industry players are forming to compete with Amazon;
- industry and category-specific predictions for how companies will remain competitive;
- analysis of how retailers are adopting IoT-enabled omnichannel marketing strategies;
- how retailers are adopting AI and machine learning to optimize inventory planning and merchandizing.

Conclusion

14　In the era of IoT, companies need to disrupt themselves, to rethink the processes and solutions that have served them well for years before their competitors do it for them. We believe that by 2035 there will be a trillion connected devices delivering orders of magnitude more data than we see today. Arm and its partners are delivering the solutions that will enable that future, a future that

---

warehouse /ˈweəhaʊs/ *n.*
a large building for storing things before they are sold, used, or sent out to shops

exploit /ˈeksplɔɪt/ *v.*
to use something in a way that helps you

incumbent /ɪnˈkʌmbənt/ *n.*
a person or business that holds a particular position in a company, market, industry, etc. at the present time

category /ˈkætɪɡərɪ/ *n.*
(in a system for dividing things according to appearance, quality, etc.) a type, or a group of things having some features that are the same

inventory /ˈɪnvəntərɪ/ *n.*
a detailed list of all the things in a place

magnitude /ˈmæɡnɪtjuːd/ *n.*
the large size or importance of something

is happening right now. I'd love to hear about your experiences with IoT and how we might work together to help you achieve your IoT data visions.

Word count: 940

Source: Patel D. The IoT data opportunity for logistics companies is here[EB/OL]. (2018-12-11)[2019-05-15]. https://community.arm.com/iot/b/internet-of-things/posts/iot-data-opportunity-for-logistics-companies-is-here.

I Reading

Task 1

Directions: Please answer the following questions.

1. When customers in the logistics world such as retailers, transportation have conversations with me, what kind of mixed feelings do they have?

2. In what way does the Pelion IoT platform satisfy the demand of customers?

3. In the logistics space, what do customers desire?

4. What has enhanced Shiseido's CRM engagement and ad performance?

5. How does Zebra manage its warehouse?

6. In the era of IoT, in order to get a competitive edge over peers, what should businesses do?

Task 2

Directions: Paraphrase the following sentences from the article; paraphrase is a restatement of a text, passage, or sentence, giving the meaning in another form. Make sure you change the underlined words in the following sentences.

1. But many companies not that far down the <u>deployment path</u> run into a <u>roadblock</u>, namely complexity. (para. 2)

2. Arm has worked <u>relentlessly</u> for years to <u>conceive</u> and deliver hardware IP solutions to support companies' IoT <u>ambitions</u>. (para. 4)

_____

_____

3. Merchandising and marketing functions in retail are <u>evolving</u> to become more effective in the omnichannel world. (para. 8)

_____

_____

II Vocabulary

Key words and expressions

logistics	intriguing	pursue	transform
commission	respondent	roadblock	vendor
myriad	at hand	enormous	daunt
scalable	navigate	dense	stitch
relentless	conceive	deliver	span
launch	enable	friction	confusion
yearn	accelerate	query	ingest
visibility	shipment	retailer	merchandise
leading-edge	cosmetics	warehouse	exploit
incumbent	category	inventory	magnitude

Terminology

Arm	n.	安谋国际科技股份有限公司
deployment path	n.	部署路径
connectivity	n.	连通性；互联互通
ERP system (enterprise resource planning system)	n.	企业资源规划系统
3PLs (third-party logistics service provider)	n.	第三方物流
untapped data	n.	未挖掘的数据
data-insight	n.	数据洞察力
ROI (return on investment)	n.	投资回报率
IP solutions	n.	IP解决方案
time to value	n.	价值实现时间
real-time data	n.	实时数据
petabytes	n.	拍字节

Gartner Magic Quadrant	n.	高德纳魔力象限
barcode scans	n.	条码扫描
treasure trove	n.	宝藏,宝库
omnichannel	n.	多渠道
McKinsey	n.	麦肯锡
Shiseido	n.	资生堂(日本公司)
loyalty membership	n.	忠实会员
public data-management platform (DMP)	n.	数据管理平台
CRM (customer relationship management)	n.	客户关系管理
Zebra Technologies	n.	斑马技术公司
load-monitoring	n.	负荷测录
ruggedized	adj.	加固抗震的

Task 1 Word bank

Directions: Please fill in the blanks with the words listed below. You may need to change the form when necessary.

logistics	intriguing	dense	deliver	accelerate
retailer	daunt	transform	visibility	cosmetics
conceive	pursue	exploit	enormous	launch

1. It cannot be denied that the reform and opening-up has _____ China.

2. The _____ of happiness has been an essential part of human progress.

3. The increasing _____ of the city's poor and homeless has forced the council into taking action.

4. Some women spend a fortune on _____ in order to appear more attractive.

5. This _____ book is both thoughtful and informative.

6. These are the reasons why we are resolved to _____ a large-scale production campaign.

7. She worked _____ hard on the project.

8. JD. com supports payment upon _____.

9. _____ system includes customer service, packaging, transportation, storage, distribution processing and information control.

10. The fabrics are sold wholesale to _____, fashion houses, and other manufacturers.

Task 2 Meaning in context

Directions: The meaning of a word may differ depending on the context. Please write down the Chinese version of the underlined phrases or words in each sentence; then translate the following paragraph into Chinese. You should pay special attention to the words in bold.

(1) develop

a. It's hard to say at this stage how the market will <u>develop</u>.

b. Their bodies were well developed and super fit.

c. Membership in her church youth group helped develop her political ideas.

d. Shipwrecks give us valuable information about how marine plants and animals develop.

e. In this way, you can extend and develop these frameworks for novel problem domains.

f. I have to develop a photograph.

g. Tobacco is a killer. Smokers and other tobacco users are more likely to develop disease and die earlier than people who don't use tobacco.

(2) business

a. His dream school is Harvard Business School.

b. The majority of small businesses fail within the first twenty-four months.

c. I'm here on business.

d. You can't mix business with pleasure.

e. They worried that German companies would lose business.

f. May I ask you what business you're in?

g. She doesn't want the police involved, as it's her business.

h. You can't stay in business without cash.

(3) connect

a. You can connect the speakers to your CD player.

b. The shower is easy to install—it needs only to be connected up to the hot and cold water supply.

c. I hoped he would not connect me with that now-embarrassing review I'd written seven years earlier.

d. In the Reagan era, well-connected Republicans received favored treatment in this organization.

e. China's aviation industry contributes to connectivity along Belt and Road.

(4) administer

a. The plan calls for the UN to administer the country until elections can be held.

b. The physician may prescribe but not administer the drug.

c. This allows you to administer your server from remote computers.

d. Judges administer justice and punishment.

(5) extract

a. Citric acid can be extracted from the juice of oranges, lemons, limes or grapefruit.

b. A dentist may decide to extract the tooth to prevent recurrent trouble.

c. He extracted a small notebook from his hip pocket.

d. Britain's trade figures can no longer be extracted from export-and-import documentation at ports.

e. Read this extract from an information booklet about the work of an airline cabin crew.

For **developers** and **business** leaders, Pelion IoT platform accelerates the time to value of IoT deployments by helping you easily **connect** trusted IoT devices on global networks, seamlessly **administer** them, and **extract** real-time data from them to drive competitive advantage. (para. 5)

Ⅲ Speaking

Directions: Please mock a product launch/press conference through role-play. The class should be divided into two teams. One team is the Arm sales team introducing the company's products and solutions, while the other team plays the role of reporters and asks questions. Then swap.

Arm side

- Host: introduces every presenter and chairs the conference
- Product and service ambassador: demonstrates cool user-experience and explains how the users benefit from the product
- Arm's data platform leader: explains what companies can do through the platform
- Pelion IoT Platform leader: explains what companies can do through the platform

The press

There are 3~4 press reporters asking questions from different aspects. For example, there will be reporters looking at:
- What challenges and opportunities will retailers be facing?
- Competitors' products and services.
- What are the strengths and drawbacks of many companies not that far down the deployment path?

Ⅳ Writing

1. What is a comparison or contrast essay

A comparison or contrast essay is an essay in which you either compare something or contrast something.

A comparison essay is an essay in which you emphasize the similarities, and a contrast essay is an essay in which you emphasize the differences.

2. There are two methods to organize a comparison and contrast essay

- Point-by-point method (or alternating arrangement)

You alternate points about A (one thing) with comparable points about B (another thing).

Example: *Where to take a vacation (mountain V.S. beach)*:

Introduction
Mountain 　A. Distance 　B. Climate 　C. Types of activities
Beach 　A. Distance 　B. Climate 　C. Types of activities
Conclusion

- Item-by-item method (or block arrangement)

You discuss all of A, then all of B.

Example: *Where to take a vacation (mountain V.S. beach)*:

Introduction
First difference between mountains and beaches is distance 　A. Mountain 　B. Beach
Second difference between mountains and beaches is climate 　A. Mountain 　B. Beach
Conclusion

3. Usually, comparison and contrast essay writing consists of three parts

- Introduction: Often begin with a sentence that will catch the reader's interest. Then name the two subjects and state that they are very similar or quite different, having many important (or interesting) similarities or differences.
- Body: Discuss how both subjects are different or similar with the two methods (alternating or block arrangement).
- Conclusion: Give a brief, general summary of the most important similarities and differences. End with a personal statement, or a prediction.

4. What do the instruction and conclusion look like

The introduction of your essay should mention both subjects and end with a strong and clearly defined thesis statement.

The conclusion of your essay should include final correlations about the two subjects and a restatement of your thesis.

5. If you want to compare and contrast, you might find the following expressions useful

Useful Expressions for Comparison and Contrast

Comparison	Contrast
• like	• unlike
• the same as	• in contrast to
• alike	• different from
• similar	• less
• likewise	• whereas
• and, as well	• however
• also, too	• but
• just as, as do, as did, as does	• as opposed to
• both	• on the other hand

Task　Essay writing

Directions: As is suggested in the article, IoT has transformed many businesses and industries, including retailing companies. It is interesting to explore what changes have taken place in the retailing industry because of IoT application. Write an essay to compare and contrast the past and present situation of the retailing industry since IoT was embraced. You should write at least 120 words.

KEY

I Reading

Task 1

1. They know there's enormous value in embracing IoT and leveraging the vast amounts of untapped data that await them to transform their business. But they are daunted by where to begin.

2. Pelion IoT platform enables organizations to connect and manage their IoT devices and data, removing much of the friction and confusion from the process of building an IoT solution.

3. In the logistics space, customers want real-time visibility on shipments with fewer barcode scans. They want to move from just tracking things from point A to point B, to capturing and tracking more data, seamlessly managing all the different data types and connecting this treasure trove of data into their existing ERP systems.

4. Shiseido, the world's third-largest cosmetics company, uses Arm's data platform to ingest, integrate, store, prepare and then discover vast amounts of data, including loyalty membership data, second-party data such as media website viewing data, and third-party public data-management platform (DMP) information.

5. Zebra Technologies' real-time load-monitoring solution SmartPack Trailer installed at warehouse doors helps optimize loading and ensure the right package ends up in the right truck.

6. In the era of IoT, companies need to disrupt themselves, to rethink the processes and solutions that have served them well for years before their competitors do it for them.

Task 2 (Tentative answers)

1. But many companies that have not fully deployed IoT meet a challenge, that is, they find IoT too complicated.

2. For years, Arm has worked persistently to develop and deliver hardware IP solutions to support companies that wish to fully apply IoT.

3. Merchandising and marketing functions in retail are developing. They are effective tools in the omnichannel retailing industry.

II Vocabulary

Task 1

1. transformed   2. pursuit   3. visibility   4. cosmetics   5. intriguing

6. launch   7. enormously   8. delivery   9. Logistics   10. retailers

Task 2

(1) develop
　a. 发展   b. 发育   c. 形成   d. 生长   e. 开发   f. 冲洗   g. 患病

(2) business
　a. 哈佛商学院   b. 小企业   c. 出差   d. 工作   e. 失去生意/业务   f. 行业   g. 私事   h. 营业

(3) connect
　a. 连接   b. 接通   c. 与……联系起来   d. 关系硬的，有人脉的   e. 互联互通

(4) administer
　a. 监管   b. 派发（药物）   c. 管理   d. 执法，执行审判

(5) extract
　a. 提炼，提取   b. 拔牙   c. 拿出   d. 摘取（信息）   e. 选段，摘录

翻译：对于开发人员和业务领导者而言，Pelion IoT 平台可帮助您轻松连接全球网络上的可信 IoT 设备，无缝管理这些设备，并从中提取实时数据以提高竞争优势，从而加快物联网部署的价值实现时间。

IoT Policy Considerations

- Social-ethical Considerations
- The Implications of the Internet of Things on Social Media and Branding
- Security and the Internet of Things
- The Internet of Things Is Revolutionizing Our Lives, but Standards Are a Must
- What the Internet of Things Means for Consumer Privacy

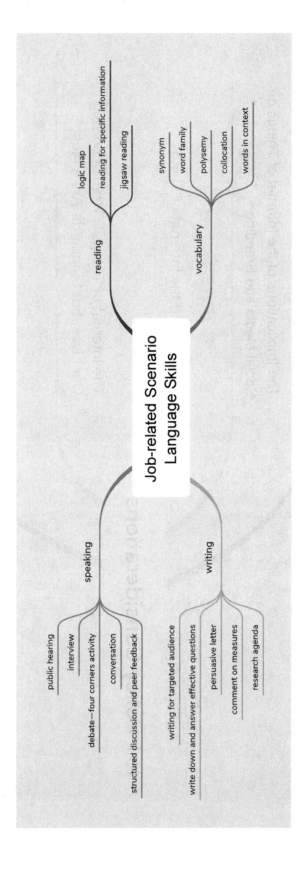

Chapter 3   IoT Policy Considerations

This chapter provides 5 articles on some IoT-related social concerns. You will be put in a job-related scenario, in which you will learn how to analyze the logic most efficiently, engage in inspiring peer feedback discussion, write for targeted audience as well as ask and answer effective questions.

Unit 11   Social-ethical Considerations

1   All new technological developments have some effect on society, whether desired or undesired. Similarly, social norms affect the direction of innovation and the diffusion of technology. Society as we know it will certainly be transformed by the diversification and widespread adoption of ubiquitous communications, RFID, **nanotechnology**, **sensor** technologies and **smart objects**. Beyond their implications for privacy protection, these technologies raise a number of other important socio-ethical issues.

2   The Internet of Things will enable levels of convenience (e.g. in the home, in the car, and in the shop) that are far ahead of current services, and are likely to have a significant positive impact on quality of life. Sensor technologies and **tagged objects** in a smart home could help care for children, the elderly and the infirm. They could also allow flexible working hours and reduced commuting times, thereby encouraging family time. But the revolutionary capacity of "anywhere, anytime, anyone and anything" communications may raise some concerns about the future of social norms and ethical values. As seen above, institutions, regulation and policy-making may not be moving fast enough to keep up with rapid technological innovation. The

---

desire /dɪˈzaɪə/ v.
to want sth.

norms /nɔːmz/ n.
ways of behaving that are considered normal in a particular society

diffusion /dɪˈfjuːʒən/ n.
the act of causing to become widely known or used

diversification
/daɪˌvɜːsɪfɪˈkeɪʃən/ n.
the act of introducing variety

the infirm /ɪnˈfɜːm/ n. (plural)
people who are weak or ill, and usually old

commute /kəˈmjuːt/ v.
travel a long distance every day between home and place of work

same can be said about individuals and society as a whole—is society evolving fast enough to adapt to these new technologies? This raises another important question—do individuals and society (like laws and regulations) have to adapt at the same pace as the technology they use?

3　One of the most common fears is that new technologies are fostering a growing atmosphere of surveillance. There are of course different forms of surveillance, but the most prevalent today are technologically mediated forms. Such forms currently include especially **video capture** in public places, government records, **biometric information** (e.g. fingerprints), **location and navigation technologies** (e.g. GPS and even mobile communications that allow the capture of basic location information). These do not necessarily occur in isolated forms, but usually work within larger systems of surveillance (e.g. security systems, governance systems). New forms, such as RFID tags, are already beginning to make their appearance on clothing to enable retailers to track product use—these may not be automatically deactivated after purchase. Similar tags and sensors on other items such as food products and household appliances open up new mechanisms for tracking and monitoring human behaviour.

4　Whether real or imaginary, an environment of surveillance can instill distrust and fear in humans, creating heightened anxiety in the exercise of choice and the making of decisions, no matter how small. Since decision-making is essential to individual self-fulfilment and self-expression, it is also crucial to societal advancement as a whole. On the other hand, suspicion and paranoia detract from healthy social intercourse, as well as creativity and overall human development. Moreover, although the creation of a network of smart things will further automate daily tasks, the complete automation of human activity is not a desirable outcome. Whether implanted with an **electronic code** or not, each human is unique and should be treated as such. There is little to gain from an increasingly conformist and uniform society. Individuality and self-expression are important catalysts for societal progress. A techno-political environment (in which individuals become mere numbers) must not be encouraged. Moreover, some numbered individuals may be included, while

---

adapt to to adapt or conform oneself to new or different conditions

surveillance /sɜːˈveɪləns/ n. careful watching of someone, especially by an organization such as the police or the army
mediated /ˈmiːdɪeɪt/ adj. acting or brought about through an intervening agency
capture /ˈkæptʃə/ n. the act of representing or recording in lasting form

deactivate /diːˈæktɪveɪt/ v. to make inactive
imaginary /ɪˈmædʒɪnərɪ/ adj. having existence only in imagination, not real
instill in v. to impress something into someone's mind
self-fulfillment /ˈselfulˈfilmənt/ n. the fulfillment of your capacities
self-expression /ˌselfɪkˈspreʃən/ n. the expression of one's individuality
paranoia /ˌpærəˈnɔɪə/ n. intense anxiety or worry
detract from to impact someone or something negatively
implant /ɪmˈplɑːnt/ v. deeply rooted; firmly fixed or held
conformist /kənˈfɔːmɪst/ adj. marked by convention and conformity to customs or rules or styles

Chapter 3　IoT Policy Considerations

arbitrary /ˈɑːbɪtrərɪ/ *adj.*
based on or subject to individual discretion or preference
clustering /ˈklʌstərɪŋ/ *n.*
a grouping of a number of similar things
breed /briːd/ *v.*
causing bad feeling or bad behaviour to develop

ephemeral /ɪˈfɛmərəl/ *adj.*
lasting a very short time

confirm /kənˈfɜːm/ *v.*
establish or strengthen as with new evidence or facts
primate /ˈpraɪmeɪt/ *n.*
a member of the group of mammals that includes humans, monkeys, and apes
detect /dɪˈtɛkt/ *v.*
discover or determine the existence, presence, or fact
lengthy /ˈlɛŋkθɪ/ *adj.*
of relatively great or tiresome extent or duration
dependability /dɪˌpɛndəˈbɪlətɪ/ *n.*
the quality of being dependable or reliable
disseminator /dɪˈsɛmɪneɪtə/ *n.*
someone or something that spreads the news
wield /wiːld/ *v.*
have and exercise

others might be excluded, in many cases in an arbitrary and automated fashion. This form of **"social sorting"** has led to the creation of an industry devoted to the clustering of various populations, breeding further discrimination and fear.

5　Furthermore, as communications between people become increasingly mediated by technology (e. g. SMS), content must not give way to form: traditional forms of intimacy and human contact must not be lost. In many instances, like the technical devices that facilitate them, human relationships are increasingly transient and ephemeral. The mobile phone is an early example of this—many people now prefer to text each other than phone each other. Many sociologists point to the increasing use of **voicemail** and **call screening**: overall, a growing resistance to human face-to-face or even vocal interaction.

6　A further implication relates to the evolution of human capabilities. As sensor technologies become increasingly sophisticated, they may replace human abilities to sense the environment. The human nose, for instance, includes a thousand sensor genes, but today's human being only uses thirty percent of these as many are no longer required (e. g. for hunting or escaping physical danger in the wild). Recent genetic studies have confirmed a decline in the number of functional **olfactory receptor genes** from primate evolution to humans. One can wonder whether the human sense of smell will evolve further, as sensors begin to replace human noses to detect fire, rotten food, and even good wine. Before the invention of writing, human beings were able to memorize lengthy texts and detailed information. Many argue that human memory, too, has diminished, as many people today cannot even remember telephone numbers due to the **memory-dialling function** of their mobile phones.

7　Further ethical considerations relate to the influence and dependability of information. As communication technologies become ubiquitous, the flow of information will be increasingly influential. Today's disseminators of information (e. g. search engines) already wield considerable power, in terms of which information is made available and in which order of priority. Often, information can be wrong, outdated, unavailable or presented out of context. Currently, these problems are limited to the World Wide Web. But as information begins to flow out from every device and thing, who will ultimately retain control over its dissemination? Care must be taken not to allow the disseminator(s) of this wealth of information to be governed by commercial interests, but rather by the public interest.

8 The emerging technologies underlying the Internet of Things offer many benefits for users and businesses alike. However, their growing complexity and availability will have a significant impact upon society. It is only through increased awareness of the far-reaching implications of these new technologies, particularly in the field of socio-ethics, that humanity itself can be preserved in an increasingly pervasive and automated technological environment.

Word count: 1,026

Source: Adapted from International Telecommunication Union. ITU Internet Report 2005: The Internet of Things [R]. Geneva: ITU,2005.

Ⅰ Reading

The ability to analyze the logic is essential to the world of work. When reading a memo or e-mail, business letter, sales report or proposal, you always want to grasp the main idea in the fastest way so that you can be efficiently informed of some important policies or information, and respond to your peers, clients or supervisors or take action accordingly. A logic map is of great help in this regard. Whatever you are reading, be it an article in the business magazine, an academic paper in your industry or any document in your workplace, try to draw a logic map that shows clearly the author's key logic markers and examples used to back up his logic. But make no mistake; there is definitely NO "standard" way to draw a logic map. Any logic map is well-drawn as long as you find it helpful. You can create your own logic map with condensed phrases and key words from the article as well as arrows and boxes, highlights and bold type to make it easier for you to read.

Task 1

Directions: Read the text and complete the following logic map. You can use this one as reference and draw your own logic map.

Chapter 3 IoT Policy Considerations | 135

Logic Map of
"Social-ethical Considerations"

⊙ **Thesis:** Technologies have both _____ and _____ **effect** on society and cause some _____ issues. (para. 1)

⊙ _____ **effect:** IoT will enable _____ (para. 2)

Examples— care for _____
allow _____, less _____, more _____

⊙ _____ **effect:** IoT raises **concerns about** _____ and _____ (para. 2)

a. New technologies are fostering _____ (para. 3)
(Examples include **current** and **new** forms of surveillance mediated by technology.)

Consequences: instill _____ in humans (para. 4)
Heighten _____ in _____
Discourage _____
Impede _____
Lead to _____
Form _____
Breed _____

b. Human relationships are increasingly _____ (para. 5)
Examples: mobile phone (prefer to _____ than _____ each other)
(Solution: _____)

c. Sensor technologies may replace _____ (para. 6)
Examples: human sense of smell (to _____)
human memory (to _____)

d. Further ethical considerations relate to _____ (para. 7)
(Solution: _____)

⊙ Summary and Appeal → People should increase their _____ of the far-reading implications of these new technologies, particularly in the field of _____ (para. 8)

Task 2 Reading for specific details

Directions: Read the text carefully and answer the following questions with key words.

1. What conveniences in our daily life can be enabled by IoT? Can you think of other specific examples?

_____

_____

2. As a result of IoT's capacity of "anywhere, anytime, anyone and anything", what

concerns do people have about the future of social norms and ethical values?

3. What are the bright and dark sides of this "growing atmosphere of surveillance" respectively?

4. In general, what possible implications do modern technologies have on human relationship and human capabilities?

II Vocabulary

Key words and expressions

desired	norms	diffusion	diversification
the infirm	commute	adapt to	surveillance
mediated	capture	deactivate	imaginary
instill in	self-fulfillment	self-expression	paranoia
detract from	implant	conformist	arbitrary
clustering	breed	ephemeral	confirm
primate	detect	lengthy	dependability
disseminator	wield		

Terminology

nanotechnology	n.	纳米技术
smart objects	n.	智能对象
tagged objects	n.	标记对象
video capture	n.	视频采集
biometric information	n.	生物信息
location and navigation technologies	n.	定位导航技术
electronic code	n.	电子编码
social sorting	n.	社会分类
voicemail	n.	语音信箱
call screening	n.	呼叫筛选
olfactory receptor genes	n.	嗅觉受体基因
memory-dialing function	n.	记忆拨号功能

Task 1

Directions: Replace the parts in bold type with appropriate words and expressions from the text.

1. Modern technologies have brought about tremendous changes in our life, whether **good or bad**.

2. About 200 new buzzwords **have found their way to** the *Oxford Dictionary* this year.

3. The Internet of Things will have **deep influence** on human society.

4. The diagram **implanted** a dangerous prejudice firmly **in** the minds of countless economics students.

5. My advice: instead of panicking, use economic breakdowns as a **driving force** for your own financial breakthroughs.

6. He talked about the country's **short-lived** unity being shattered by the defeat.

Task 2 Word family

Directions: One way to figure out the meaning of an unknown word is to look for its relationship with other words in the same word family. Even if you cannot figure out the exact meaning, your understanding can be enough to allow you to read on. Please read the phrases from the text. Write down at least one other word you know that is related to the underlined word and look up more related words in the dictionary. An example has been given to you.

1. the diffusion of technology	diffuse; diffuser; diffused
2. enable levels of convenience	
3. societal advancement	
4. in an arbitrary and automated fashion	
5. a growing resistance	
6. a desirable outcome	
7. to sense the environment	
8. to detect fire	
9. disseminators of information	
10. memorize lengthy texts	

III Speaking

Group discussion and peer feedback are pivotal to learning new ideas, inspiring deeper thinking, sharing best practices, brainstorming creative solutions as well as strengthening effective communication and listening skills.

Task 1 Sharing comments

Directions: Work in pairs and take turns to comment on any one of the following statements.

1. Every time there's a new tool, whether it's Internet or cell phones or anything else, all these things can be used for good or evil. Technology is neutral; it depends on how it's used.

—Rick Smolan

2. Technology is nothing. What's important is that you have a faith in people, that they're basically good and smart, and if you give them tools, they'll do wonderful things with them.

—Steve Jobs

3. The first rule of any technology used in a business is that automation applied to an efficient operation will magnify the efficiency. The second is that automation applied to an inefficient operation will magnify the inefficiency.

—Bill Gates

4. I have this really bad habit of doing things on the Internet and forgetting that the whole world is going to see it.

—Maisie Williams

5. The Internet is not just one thing, it's a collection of things—of numerous communications networks that all speak the same digital language.

—James H. Clark

6. The Internet of Things tell us that a lot of computer-enabled appliances and devices are going to become part of this system, too: appliances that you use around the house, that you use in your office, that you carry around with yourself or in the car. That's the Internet of Things that's coming.

—Vint Cerf

Task 2 Structured discussions

Directions: Work in groups of 3 to 4 and review the reading material "Social-ethical Considerations". Each person highlights at least one significant idea and/or quote in the text.

Within your group:

a. Identify a facilitator/time keeper.

b. The first person presents the idea or quote they highlighted and why they think it is important. (3 minutes)

c. Each person (1 minute) responds to what they heard from the presenter and the presenter does not talk and only takes notes.

d. The presenter is given one more minute to respond to what they heard or clarify their thinking. (1 minute)

e. Take turns to be the presenter.

Ⅳ Writing

Effective writing on the job is carefully designed and clearly presented. You are always writing (1) for a targeted audience, (2) with a clearly expressed purpose, (3) about a topic your audience is interested in and needs to understand, (4) in language suitable for the specific scenario.

Task 1 Writing for a targeted audience

Directions: Though IoT helps people and the communities of which they are a part by making their lives easier and safer, there are real challenges in deploying it as mentioned in the text. With specific audience in your mind, write down **one of IoT's negative implications that may most concern your targeted audience** *and come up with some possible solutions.*

Task 2 Experience writing

Directions: Once fearful and dependent, hopeful and distrustful, our contemporary relationship with technology is highly ambivalent. Remember a time when you experienced the "ambivalence of modern technology" (Internet, IoT, RFID, nanotechnology, sensor technologies, smart objects, etc.), a time when you had mixed feelings towards technology. Write down as much as you can about that memory.

KEY

Ⅰ Reading

Task 1

Logic Map of
"Social-ethical Considerations"

⊙ **Thesis**: Technologies have both **desired and undesired effect** on society and **cause some socio-ethical issues**. (para. 1)

⊙ **Desired effect**: IoT will enable **levels of convenience** (para. 2)

Examples— care for children, the elderly and the infirm

allow flexible working hours, less commuting time, more family time

⊙ **Undesired effect**: IoT raises **concerns about social norms and ethical values**

(para. 2)

a. New technologies are fostering a growing atmosphere of surveillance (para. 3)

(Examples include **current** and **new** forms of surveillance mediated by technology.)

video capture   RFID tags on clothing   tags and sensors on food products or household appliances

Consequences: instill distrust/fear in humans (para. 4)

Heighten anxiety in decision-making

Discourage individuality and self-expression

Impede healthy social intercourse, creativity and human development

Lead to "social sorting"

Form clustering of various populations

Breed further discrimination and fear

b. Human relationships are increasingly transient and ephemeral. (para. 5)

Examples: mobile phone (prefer to text than phone each other)

(Solution: traditional forms of intimacy and human contact must not be lost.)

c. Sensor technologies may replace human abilities to sense the environment. (para. 6)

Examples: human sense of smell (to detect fire, rotten food and good wine)

human memory (to remember phone numbers)

d. Further ethical considerations relate to the influence and dependability of information. (para. 7)

Chapter 3 IoT Policy Considerations | 141

(Solution: Care must be taken not to allow the disseminator(s) of information to be governed by commercial interests, but rather by the public interest.)

⊙ **Summary and Appeal** → People should increase their awareness of the far-reading implications of these new technologies, particularly in the field of socio-ethics   (para. 8)

Task 2

1. The Internet of Things will enable levels of convenience (e.g. in the home, in the car, and in the shop) that are far ahead of current services, and are likely to have a significant positive impact on quality of life. Sensor technologies and tagged objects in a smart home could help care for children, the elderly and the infirm. They could also allow flexible working hours and reduced commuting times, thereby encouraging family time.

2. a. New technologies are fostering a growing atmosphere of surveillance.
   b. Human relationships are increasingly transient and ephemeral.
   c. Sensor technologies may replace human abilities to sense the environment.
   d. Further ethical considerations relate to the influence and dependability of information.

3. The bright side may include personal security (video capture in the public, location and navigation technologies, etc.), quality supervision of consumers goods, food products, household appliances, etc.

The dark side may include instilling distrust/fear in human beings, heightening anxiety in decision-making, discouraging individuality and self-expression, impeding healthy social intercourse, creativity and human development, leading to "social sorting, forming clustering of various populations and breeding further discrimination and fear, etc.

4. a. Human relationships are increasingly transient and ephemeral. For example, people prefer to text than phone each.
   b. Sensor technologies may replace human abilities to sense the environment. For example, human sense of smell (to detect fire, rotten food and good wine) and human memory (to remember phone numbers) have been gradually replaced by sensor technologies.

Ⅱ Vocabulary

Task 1

1. whether desired or undesired
2. make their appearance on
3. far-reaching implications
4. instilled…in
5. catalyst
6. ephemeral

Task 2

1. diffuse; diffuser; diffused
2. enabling; enabler; enablement; enabled
3. society; social; socialist; socialized; socialistic
4. automate; automatic; automation; automatically; automatism
5. resist; resistible; resistant; resistive; resistless
6. desire; desired; undesired; desirous; desirability; desirableness
7. sensible; sensing; sensory; sensation; sensational; senseless; sensibility
8. detective; detection; detector; detectable
9. disseminate; dissemination; disseminative
10. memory; memorial; memorable; memorizer

Unit 12 The Implications of the Internet of Things on Social Media and Branding

1 Soon, you won't be reading about "The Internet of Things" because it really won't be a separate topic, it will infiltrate our way of life and become a commonality. However, until that happens, we are getting used to the idea of objects being connected to the Internet and every "thing" becoming a SMART thing. As companies and their software and technology evolve into this new connected reality, the marketing departments for those companies will also have adjustments to make, especially in the areas of social media and branding.

Smart consumers will love smart technology

2 Since objects will become "SMART" objects, the potential is there for them to take on a life of their own and consumers will need to be educated in what that means for them and their relationship with your company. The marketing departments will be tasked with communicating this information to consumers in a way that they can not only understand, but adopt. There is bound to be resistance from consumers who are concerned about privacy. Granted, some might even feel paranoid about the tracking and automated communications that are implied when objects start communicating with each other and building data. However, we keep seeing glimpses of the death of privacy, or at the very least, the end of privacy as we know it. Large portions of our population already share intimate details about their lives on social media. Most consumers are also already willing to give their personal data online in exchange for discounts or special offers, or even customization. This is referred to as **conscious privacy**. When objects become smart and are connected to social life, they will be "able to communicate with one another and 'connect' to other people or devices in the same vicinity based on the behavior of the person".

The time is now!

3 The success of your marketing and the incorporation of the

---

branding /ˈbrændɪŋ/ n.
the activity of giving a particular name and image to goods and services so that people will be attracted to them and want to buy
infiltrate /ˈɪnfɪlˌtreɪt/ v.
permeate or become a part of (something) in this way
commonality /ˌkɒməˈnælɪti/ n.
a shared feature or attribute
department /dɪˈpɑːtmənt/ n.
a section of a large organization such as a government, business, university etc.

granted /ˈɡrɑːntɪd/ adv.
used to say that you accept that something is true, often before you make another statement about it
paranoid /ˈpærənɔɪd/ adj.
afraid or suspicious of other people and believing that they are trying to harm you, in a way that is not reasonable
glimpse /ɡlɪmps/ n.
a faint and transient appearance, an occasionally perceptible resemblance; a tinge or trace (of a quality)
offer /ˈɒfə/ n.
a reduction in the normal price of something, usually for a short period of time
customization /ˌkʌstəmaɪˈzeɪʃən/ n.
the action of making or changing something according to the buyer's or user's needs
vicinity /vɪˈsɪnɪti/ n.
the area around a particular place

Internet of Things will depend on how you disseminate the information that will help your consumers understand the <u>ins and outs</u> of the technology and how you will use it. The time to start messaging and educating your public about the Internet of Things is now. Do not wait until your company has fully integrated new technology to start explaining or announcing change to your customers. Start writing about how your company's plans to upgrade software and technology and how IoT will impact your company, or more importantly, how it will impact your customers. Depending on how much your company will integrate IoT in the future, will it impact your brand, your daily operations, or even your **mission statement**? Will your branding strategy need to be modified? Your social channels and blog should start regularly producing content education for your customers and slowly making the switch to smart technology. A gradual change is easier for most to accept rather than a sudden change, especially where technology is concerned. If your technology upgrades will mean that there is a **learning curve** for your customers, then you can begin to help them <u>acclimate</u> through your content and your social messaging.

Examples of social media with the Internet of Things

4 There are already companies that are doing some amazing marketing and social media campaigns that incorporate the Internet of Things and this new technology. Social Media Today highlights some of this technology that is paving the way for the future. Examples include smart <u>vending machines</u>, video media advertising, in-store retail mobile applications, and much more.

Action items and takeaways for your company

5 The messaging about the Internet of Things and your technology upgrades should <u>encompass</u> all of the channels you currently use to communicate with your customers and possibly <u>tap</u> some new ones. In addition to social media messaging and building out educational content on your company blog, you can implement educational content in the follow ways to educate your consumers:

- **Webinars**
- Videos

---

ins and outs the intricate details of a situation or process

acclimate /ˈæklɪˌmeɪt, əˈklaɪmɪt/ *v.* = acclimatize, to get used to a new place, situation or climate

vending machine a machine from which you can buy cigarettes, drinks, etc. by putting coins into it

encompass /ɪnˈkʌmpəs/ *v.* to include a large number or range of things

tap /tæp/ *v.* to make use of a source of energy, knowledge, etc. that already exists

- Online presentations
- **White papers**

6 Try to cover all your <u>bases</u> so that your audience receives the message enough times to understand its importance and has the option to learn about it in their preferred format. <u>Address</u> privacy concerns and explain how your company will be handling private data and <u>personalization</u>. Those are bound to be the top two concerns that consumers will have when it comes to the Internet of Things. Explain how your company will benefit from the added convenience of new technology and <u>upgrades</u> and then also explain how you will pass on your savings to your customers. <u>Detail</u> your company's **opt out policy** so that your customers will understand that they can be in control of their own data. Lastly, <u>monitor</u> what consumers and the media are saying about your company so that you can understand your customers' perspectives even if they aren't providing you with feedback.

Word count: 802

Source: Victor E. The implications of the Internet of Things on social and branding [EB/OL]. (2015-05-09) [2020-05-15]. https://www.socialmediatoday.com/technology-data/elizabethvictor/2015-05-28/internet-things-and-implications-social-and-branding.

---

base /beɪs/ *n.*
an idea, a fact, a situation, etc. from which something is developed

address /əˈdres/ *v.*
to think about a problem or a situation and decide how you are going to deal with it

personalization *n.*
the process of making something suitable for the needs of a particular person

upgrade /ˈʌpɡreɪd/ *n.*
a more powerful and efficient piece of machinery, computer systems, etc

detail /ˈdiːteɪl/ *v.*
to give a list of facts or all the available information about something

monitor /ˈmɒnɪtə/ *v.*
to watch and check something over a period of time in order to see how it develops, so that you can make any necessary changes

Ⅰ Reading

The ability to analyze the logic is essential to the world of work. When reading a memo or e-mail, business letter, sales report or proposal, you always want to grasp the main idea in the fastest way so that you can be efficiently informed of some important policies or information, and respond to your peers, clients or supervisors or take action accordingly. A logic map is of great help in this regard. Whatever you are reading, be it an article in the business magazine, an academic paper in your industry or any document in your workplace, try to draw a logic map that shows clearly the author's key logic markers and examples used to back up his logic. But make no mistake; there is definitely NO "standard" way to draw a logic map. Any logic map is well-drawn as long as you find it helpful. You can create your own logic map with condensed phrases and key words from the article as well as arrows and boxes, highlights and bold type to make it easier for you to read.

Task 1

Directions:Read the text and complete the following logic map. You can use this one as reference and draw your own logic map.

Logic Map of
"The Implications of the Internet of Things on Social Media and Branding"

⊙ **Thesis**: As "The Internet of Things" will _____ our way of life and become a _____, companies will have adjustments to make, especially in the areas of _____ and _____. (para. 1)

⊙ **Reasons for adjustment**: Smart consumers will love _____ despite privacy _____. (para. 2)

a. **Some consumers** might feel _____ about the tracking and automated communications. (para. 2)

b. **Most consumers** already share _____ details about their lives on _____. Most consumers are also already willing to give their _____ online in exchange for _____ or special _____, or even _____. (para. 2)

Suggestions: The _____ departments will be tasked with communicating this information to consumers in a way that they can not only understand, but adopt. (para. 2)

⊙ **When to make adjustment**: The time is now. (para. 3)

Suggestion 1: Start to explain or announce changes to your customers before the company has _____. (para. 3)

Suggestion 2: Start to write about company's plans to upgrade _____ and _____ and explain how IoT will impact the company (_____, _____, or even _____) and customers. (para. 3)

Suggestion 3: Change _____ rather than suddenly and begin to help customers _____ through the _____ and the social messaging. (para. 3)

Successful Examples of Social Media with IoT: _____, _____, _____, etc. (para. 4)

⊙ **What actions to take?**

Using all the channels including _____, _____, _____, _____, _____, and possibly tap some new ones, so as to communicate with and educate customers. (para. 5)

Suggestions for actions to take (para. 6)

a. Cover all your _____. (para. 6)

b. Address _____ and explain how your company will be handling _____ and _____. (para. 6)

c. Address top two consumer concerns: explain how your company will benefit from the _____ of IoT and how you will pass _____ to customers. (para. 6)

d. Detail your company's _____ policy. (para. 6)

e. Monitor what consumers and the media are saying about your company. (para. 6)

Task 2　Reading for specific details

Directions: Read the text carefully and answer the following questions with key words.

1. How has IoT infiltrated our way of life and become a commonality? Can you think of specific examples?

2. How do you understand "conscious privacy"?

3. What measures should companies take in order to succeed in marketing and incorporating IoT?

4. What are the top two concerns that consumers will have when it comes to the IoT?

II Vocabulary

Key words and expressions

branding	infiltrate	commonality
department	granted	paranoid
glimpse	offer	customization
vicinity	ins and outs	acclimate
vending machine	encompass	tap
base	address	personalization
upgrade	detail	monitor

Terminology

conscious privacy	n.	有意识提供隐私信息的行为
mission statement	n.	宗旨声明
learning curve	n.	学习曲线
Social Media Today	n.	《今日社交媒体》
webinars	n.	网络研讨会
white papers	n.	白皮书
opt out policy	n.	退出政策

Task 1

Directions: Put appropriate prepositions or adverbs in the blanks.

1. As companies and their software and technology evolve _____ this new connected reality, the marketing departments for those companies will also have adjustments to make.

2. There is bound to be resistance _____ consumers who are concerned about privacy.

3. Most consumers are also already willing to give their personal data online in exchange _____ discounts or special offers, or even customization.

4. Your social channels and blog should start regularly producing content education for your customers and slowly making the switch _____ smart technology.

5. Explain how your company will benefit from the added convenience of new technology and upgrades and then also explain how you will pass _____ your savings to your customers.

6. Detail your company's opt out policy so that your customers will understand that they can be _____ control of their own data.

7. Since objects will become "SMART" objects, the potential is there for them to take _____ a life of their own and consumers will need to be educated in what that means for them and their relationship with your company.

8. Granted, some might even feel paranoid _____ the tracking and automated communications that are implied when objects start communicating with each other and building data.

Task 2

Directions: Translate the following sentences and phrases, paying special attention to the different meanings of the words in bold type.

1. turn off the **tap**; a **tap** on the door; **tap** your fingers; **tap** telephones

2. the **details** of plan; the report **detailed** policy changes; see the bottom of this page for **details**; a **detail** of four men

3. What's your **address**; website **address**; to **address** a conference; schedule an **address**; He always **addresses** me as "sir".

4. a special **grant**; **grant** more independence to the children; take something for **granted**

5. **offer** to help; reconsider her **offer**; **offer** advice on investment; the **offers** in the shop

Task 3　Word family

Directions: One way to figure out the meaning of an unknown word is to look for its relationship with other words in the same word family. Even if you cannot figure out the exact meaning, your understanding can be enough to allow you to read on. Please look at the phrases from the text. Write down at least one other word you know that is related to the underlined word and look up more related words in the dictionary. An example has been given to you.

1. become a <u>commonality</u>	common; commoner; commonage; commonly
2. have <u>adjustments</u> to make	
3. <u>conscious</u> privacy	
4. or even <u>customization</u>	
5. help them <u>acclimate</u>	
6. private data and <u>personalization</u>	
7. share <u>intimate</u> details	

III Speaking

Team discussion and peer feedback are pivotal to learning new ideas, inspiring deeper thinking, sharing best practices, brainstorming creative solutions as well as strengthening effective communication and listening skills.

Task 1

Directions: Give a 2-minute talk on one of the following topics.

1. What are the challenges and opportunities for the public and business sector when the Internet of Things infiltrates our way of life and becomes a commonality?

2. How do you understand "conscious privacy"? Have you ever seen examples like that in your life?

3. As customers, what do you care the most about the company's incorporation of the Internet of Things?

Task 2 Structured discussions

Directions: Work in groups of 3 to 4 and review the reading material "The Implications of the Internet of Things on Social Media and Branding". Each person highlights at least one significant idea and/or quote in the text.

And then within your group:

a. Identify a facilitator/time keeper.

b. The first person presents the idea or quote they highlighted and why they think it is important. (3 minutes)

c. Each person (1 minute) responds to what they heard from the presenter and the presenter does not talk and only takes notes.

d. The presenter is given one more minute to respond to what they heard or clarify their thinking. (1 minute)

e. Take turns to be the presenter.

IV Writing

Asking good questions to each other is an important communication skill for the success of you and your team in the future workplace. In order to ask effective questions that elicit thinking, you need to:

a. Use the approachable voice;
b. Employ plural forms;
c. Use tentative/exploratory language;
d. Embed positive presuppositions;
e. Design open-ended and probing questions.

Task 1

Directions: Work in pairs. First, work on your own and write down as many effective questions related to the text as possible, and then take turns to ask and answer each other's questions.

_____
_____
_____
_____
_____
_____
_____
_____

Task 2

Directions: Choose one of the effective questions that you've designed or heard or are listed as below and write down your thoughts on it.

Approach	Possible Effective Questions
Here and now	If you were the CEO of the company and your customers who were concerned about the privacy issue caused by IoT were right here now, what would you say to them?
Future projections	What do you think the effects of IoT will have on the business branding strategies 10 years later?

_____
_____
_____
_____
_____
_____
_____

KEY

I Reading

Task 1

Logic Map of
"The Implications of the Internet of Things on Social Media and Branding"

⊙ **Thesis:** As "The Internet of Things" will <u>infiltrate</u> our way of life and become a <u>commonality</u>, companies will have adjustments to make, especially in the areas of <u>social media</u> and <u>branding</u>. (para. 1)

⊙ **Reasons for adjustment: Smart consumers will love smart technology despite privacy concerns.** (para. 2)

a. **Some consumers** might feel <u>paranoid</u> about the tracking and automated communications. (para. 2)

b. **Most consumers** already share <u>intimate</u> details about their lives on <u>social media</u>. Most consumers are also already willing to give their <u>personal data</u> online in exchange for <u>discounts</u> or <u>special offers</u>, or even <u>customization</u>. (para. 2)

Suggestions: The <u>marketing</u> departments will be tasked with communicating this information to consumers in a way that they can not only understand, but adopt. (para. 2)

⊙ **When to make adjustment: The time is now.** (para. 3)

Suggestion 1: Start to explain or announce changes to your customers before the company has <u>fully integrated new technology.</u> (para. 3)

Suggestion 2: Start to write about company's plans to upgrade <u>software</u> and <u>technology</u> and explain how IoT will impact the company (<u>branding strategy</u>, <u>daily operations</u>, or even <u>mission statement</u>) and customers. (para. 3)

Suggestion 3: Change <u>gradually</u> rather than suddenly and begin to help customers <u>acclimate</u> through the <u>content</u> and the social messaging. (para. 3)

Successful Examples of Social Media with IoT: <u>smart vending machines</u>, <u>video media advertising</u>, <u>in-store retail mobile applications</u>, etc. (para. 4)

⊙ **What actions to take?**

Using all the channels including <u>company blog</u>, <u>webinars</u>, <u>videos</u>, <u>online presentations</u>, <u>white papers</u>, and possibly tap some new ones, so as to communicate with and educate customers. (para. 5)

Suggestions for actions to take (para. 6)

a. Cover all your <u>bases</u>. (para. 6)

b. Address privacy concerns and explain how your company will be handling private data and personalization. (para. 6)

c. Address top two consumer concerns: explain how your company will benefit from the added convenience of IoT and how you will pass company's savings to customers. (para. 6)

d. Detail your company's opt out policy. (para. 6)

e. Monitor what consumers and the media are saying about your company. (para. 6)

Ⅱ Vocabulary

Task 1

1. into 2. from 3. for 4. to 5. on 6. in 7. on 8. about

Task 2

1. 关上**水龙头**；轻轻**敲**门；用手指**打拍子**；**窃听**电话

2. **计划**的**细节**；报告**详细介绍**了政策变化；**详情**请见本页末；一支四人的**分遣队**

3. 你的**地址**是；**网址**；在会议上**发言**；安排做一个**演讲**；他一直**称**我为先生

4. 特别**补助金**；**准许**孩子有更多的独立；把某件事**当成理所当然**

5. **表示愿意帮助**；重新考虑她的**提议**；**提供**投资建议；商店的**特价品**

Task 3

1. common; commoner; commonage; commonly

2. adjust; adjustable; adjusted; adjustive; adjustor; adjustability

3. consciousness; consciously

4. customer; customize; customary; custom

5. acclimation; acclimatization; acclimatize; climate

6. person; personal; personalize; personally; personality; personalized; personage

7. intimacy; intimation; intimately

Unit 13 Security and the Internet of Things

1 Nowadays, we no longer have things with computers embedded in them. We have computers with things attached to them.

2 Your modern refrigerator is a computer that keeps things cold. Your oven, similarly, is a computer that makes things hot. An ATM is a computer with money inside. Your car is no longer a mechanical device with some computers inside; it's a computer with four wheels and an engine. Actually, it's a <u>distributed</u> system of over 100 computers with four wheels and an engine.

3 The Internet is no longer a web that we connect to. Instead, it's a <u>computerized</u>, networked, and interconnected world that we live in. This is the future, and what we're calling the Internet of Things.

4 Broadly speaking, the Internet of Things has three parts. There are the sensors that collect data about us and our environment, the "smarts" that figure out what the data means and what to do about it, and the **actuators** that affect our environment. The point of a smart <u>thermostat</u> isn't to record the temperature; it's to control the <u>furnace</u> and the air conditioner. Driverless cars collect data about the road and the environment to steer themselves safely to their destinations.

5 You can think of the sensors as the eyes and ears of the Internet. You can think of the actuators as the hands and feet of the Internet. And you can think of the stuff in the middle as the brain. We are building an Internet that senses, thinks, and acts.

6 This is the classic definition of a robot. We're building a world-size robot, and we don't even realize it.

7 As all computers are <u>hackable</u>, so is this world-size robot. But until now we've largely left computer security to the market. Because the computer and network products we buy and use are so <u>lousy</u>, an enormous after-market industry in computer security has emerged. Governments, companies, and people buy the security they think they need to secure themselves. We've <u>muddled through</u> well enough, but the market failures <u>inherent</u> in trying to secure this world-size robot will soon become too big to ignore.

---

distributed /dɪˈstrɪbjuːtɪd/ *adj.*
spread out or scattered about or divided up

computerized /kəmˈpjuːtəraɪzd/ *adj.*
stored, processed, analyzed, or generated by computer

thermostat /ˈθɜːməˌstæt/ *n.*
a regulator for automatically regulating temperature by starting or stopping the supply of heat

furnace /ˈfɜːnɪs/ *n.*
a container or enclosed space in which a very hot fire is made, for example, to melt metal, burn rubbish, or produce heat for a building or house

hackable /ˈhækəbl/ *adj.*
having the risk of being broken into the system, especially in order to get secret information

lousy /ˈlaʊzi/ *adj.*
of very bad quality

muddle through manage to do something even though you do not have the proper equipment or do not really know how to do it

inherent /ɪnˈhɪərənt/ *adj.*
in the nature of something though not readily apparent

externality /ˌɛkstɜːˈnælɪtɪ/ n.
the state or condition of being external
offline /ˈɒfˌlaɪn/ adv.
not connected to the Internet
counterbalancing /ˌkaʊntəˈbælənsɪŋ/ adj. serving to equilibrate

subvert /səbˈvɜːt/ v.
destroy completely

regulatory /ˈrɛɡjʊˌleɪtərɪ/ adj.
restricting according to rules or principles
mismatch /ˈmɪsmætʃ/ n.
a bad or unsuitable match

deceptive /dɪˈsɛptɪv/ adj.
designed to deceive or mislead either deliberately or inadvertently
jurisdiction /ˌdʒʊərɪsˈdɪkʃən/ n.
the territory within which power can be exercised

horizontally /ˌhɒrɪˈzɒntəlɪ/ adv.
parallel to or in the plane of the horizon or a base line

8　Markets alone can't solve our security problems. Markets are motivated by profit and short-term goals at the expense of society. They can't solve collective-action problems. They won't be able to deal with economic <u>externalities</u>, like the vulnerabilities in **DVRs** that resulted in Twitter going <u>offline</u>. And we need a <u>counterbalancing</u> force to corporate power.

9　This all points to policy. While the details of any computer-security system are technical, getting the technologies broadly deployed is a problem that spans law, economics, psychology, and sociology. And getting the policy right is just as important as getting the technology right because, for Internet security to work, law and technology have to work together. This is probably the most important lesson of **Edward Snowden**'s **NSA** disclosures. We already knew that technology could <u>subvert</u> law. Snowden demonstrated that law could also subvert technology. Both fail unless each work. It's not enough to just let technology do its thing.

10　Any policy changes to secure this world-size robot will mean significant government regulation. And I have a proposal: a new government <u>regulatory</u> agency.

11　We have a practical problem when it comes to Internet regulation. There's no government structure to tackle this at a systemic level. Instead, there's a fundamental <u>mismatch</u> between the way government works and the way this technology works that makes dealing with this problem impossible at the moment.

12　In the U. S., the **FAA** regulates aircraft. The **NHTSA** regulates cars. The **FDA** regulates medical devices. The **FCC** regulates communications devices. The **FTC** protects consumers in the face of "unfair" or "<u>deceptive</u>" trade practices. Even worse, who regulates data can depend on how it is used. If data is used to influence a voter, it's the **Federal Election Commission**'s <u>jurisdiction</u>. If that same data is used to influence a consumer, it's the FTC's. Use those same technologies in a school, and the Department of Education is now in charge. Robotics will have its own set of problems, and no one is sure how that is going to be regulated. Each agency has a different approach and different rules. They have no expertise in these new issues, and they are not quick to expand their authority for all sorts of reasons.

13　Compare that with the Internet. The Internet is a system of integrated objects and networks. It grows <u>horizontally</u>, demolishing old technological barriers so that people and systems that never previously communicated now can. Already, apps on a smartphone can log health information, control your energy use, and communicate with your car. That's a set of functions that crosses jurisdictions of at least four different government agencies.

14   Our world-size robot needs to be viewed as a single entity with millions of components interacting with each other. Any solutions here need to be holistic. They need to work everywhere, for everything. Whether we're talking about cars, drones, or phones, they're all computers.

15   I don't think any of us can predict the totality of the regulations we need to ensure the safety of this world, but here's a few. We need government to ensure companies follow good security practices: testing, patching, secure defaults— and we need to be able to hold companies liable when they fail to do these things. We need government to mandate strong personal data protections, and limitations on data collection and use. We need to ensure that responsible security research is legal and well-funded. We need to enforce transparency in design, some sort of code **escrow** in case a company goes out of business, and interoperability between devices of different manufacturers, to counterbalance the monopolistic effects of interconnected technologies. Individuals need the right to take their data with them. And Internet-enabled devices should retain some minimal function if disconnected from the Internet.

16   The world-size robot is less designed than created. It's coming without any forethought or architecting or planning; most of us are completely unaware of what we're building. In fact, I am not convinced we can actually design any of this. When we try to design complex sociotechnical systems like this, we are regularly surprised by their emergent properties. The best we can do is observe and channel these properties as best we can. Market thinking sometimes makes us lose sight of the human choices and autonomy at stake. Before we "get controlled" by the world-size robot, we need to rebuild confidence in our collective governance institutions.

Word count: 1,085

Source: Adapted from Schneier B. With the Internet of Things, we're building a world-size robot. How are we going to control it? [EB/OL]. (2017-01) [2019-05-05]. http://nymag.com/intelligencer/2017/01/the-internet-of-things-dangerous-future-bruce-schneier.html.

---

demolish /dɪˈmɒlɪʃ/ v.
destroy completely

log /lɒg/ v.
record an event or fact officially in writing or on a computer

holistic /həʊˈlɪstɪk/ adj.
emphasizing the organic or functional relation between parts and the whole

totality /təʊˈtælɪtɪ/ n.
the state of being total and complete

patching /ˈpætʃɪŋ/ n.
computing to correct or improve by adding a small set of instruction

default /ˈdiːfɔːlt/ n.
a particular set of instructions which the computer always uses unless the person using the computer gives other instructions

liable /ˈlaɪəbəl/ adj.
held legally responsible

mandate /ˈmændeɪt/ v.
to make mandatory, as by law

transparency /trænsˈpærənsɪ/ n.
the quality of being clear and transparent

monopolistic /məˌnɒpəˈlɪstɪk/ adj. having exclusive control over a commercial activity by possession or legal grant

forethought /ˈfɔːθɔːt/ n.
planning or plotting in advance of acting

Ⅰ Reading

The ability to analyze the logic is essential to the world of work. When reading a memo or e-mail, business letter, sales report or proposal, you always want to grasp the main idea in the fastest way so that you can be efficiently informed of some important policies or information, and respond to your peers, clients or supervisors or take action accordingly. A logic map is of great help in this regard. Whatever you are reading, be it an article in the business magazine, an academic paper in your industry or any document in your workplace, try to draw a logic map that shows clearly the author's key logic markers and examples used to back up his logic. But make no mistake; there is definitely NO "standard" way to draw a logic map. Any logic map is well-drawn as long as you find it helpful. You can create your own logic map with condensed phrases and key words from the article as well as arrows and boxes, highlights and bold type to make it easier for you to read.

Task 1

Directions: Read the text and complete the following logic map. You can use this one as reference and draw your own logic map.

Logic Map of
"Security and the Internet of Things"

⊙ **What is changed by IoT:** We have computers with _____ attached to them.

(para. 1)

Examples:
- refrigerator—a computer that _____.
- oven —a computer that _____.
- ATM—a computer with _____.
- car—a computer with _____.

(para. 2)

⊙ **What is IoT:** (para. 3~6)

a. IoT is a _____ world that we live in. (para. 3)

b. IoT is a _____ with three parts:
- the _____—eyes of ears of the Internet
- the "_____"—brain of the Internet
- the _____—hands and feet of the Internet

(para. 4~6)

⊙ **What is the security concern of IoT:**

a. IoT is _____ just like computers. (para. 7)

b. _____ alone can't solve our security problems as they are motivated by _____ and _____ at the expense of society. (para. 8)

⊙ **What are the general solutions:**

a. We should get the _____ right. (para. 9)

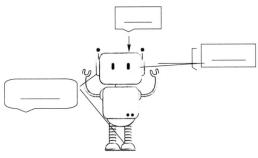

b. As policy change means significant government regulation, a new _____ is needed.

(para. 10)

● What's wrong with the current Internet regulation:

a. There is no government structure to tackle the security hazard at a _____ level.

(para. 11)

b. Each agency has a _____ approach and _____ rules.　　(para. 12)

Therefore

IoT needs to be viewed as a _____ and any solutions here need to be _____.

(para. 13~14)

● What should the government do: (para. 15)

Government needs to
- ensure companies _____ and hold them _____.
- mandate strong _____ protections and limitations on _____.
- legalize and fund _____.
- ensure _____ and _____.

● A summary:

IoT is less designed than created and we need to rebuild confidence in _____.

(para. 16)

Task 2　Reading for specific details

Directions: Read the text carefully and answer the following questions with key words.

1. Why can't we leave IoT security to the market?

_____

_____

2. What are the challenges confronting government agencies when dealing with IoT security issues?

_____

_____

3. Why do solutions to the IoT security issues need to be holistic?

4. What measures do we expect the government to take in order to ensure the safety of this world?

II Vocabulary

Key words and expressions

distributed	computerized	thermostat	furnace
hackable	lousy	muddle through	inherent
externality	offline	counterbalancing	subvert
regulatory	mismatch	deceptive	jurisdiction
horizontally	demolish	log	holistic
totality	patching	default	liable
mandate	transparency	monopolistic	forethought

Terminology

actuator	n.	传动装置
DVR(digital video recorder)	n.	硬盘录像机
Edward Snowden	n.	爱德华·斯诺登
NSA (National Security Agency)	n.	美国国家安全局
FAA(Federal Aviation Administration)	n.	美国联邦航空管理局
NHTSA(National Highway Traffic Safety Administration)	n.	美国高速公路安全管理局
FDA(Food and Drug Administration)	n.	美国食品及药品监督管理局
FCC(Federal Communications Commission)	n.	美国联邦通信委员会
FTC(Federal Trade Commission)	n.	美国联邦贸易委员会
FEC(Federal Election Commission)	n.	美国联邦选举委员会
escrow	n.	委托付款服务

Task 1

Directions: Put appropriate prepositions or adverbs in the blanks.

1. We no longer have things with computers embedded _____ them. We have computers with things attached _____ them.

2. But until now we've largely left computer security _____ the market.

3. We've muddled _____ well enough, but the market failures inherent in trying to secure this world-size robot will soon become too big to ignore.

4. There is a fundamental mismatch _____ the way government works _____ the way this technology works.

5. We need to enforce transparency in design in case a company goes _____ _____ business.

6. Market thinking sometimes makes us lose sight _____ the human choices and autonomy _____ stake.

7. We need to rebuild confidence _____ our collective governance institutions.

Task 2

Directions: Translate the following sentences and phrases, paying special attention to the different meanings of the words in bold type.

1. I was dressed in a **smart** navy blue suit; **smart** dinner parties; my eyes **smarted** from the smoke; **smart money**; **street smart**; **book smart**

2. **stuff** my hands in the pocket; **stuff** the bag full; **stuff** myself with 5 hamburgers; **stuffed** birds; **stuffed** toys; my nose is **stuffed**

3. business **failures**; engine **failure**; the marriage is a **failure**; his **failure** to pass the test; he sees himself as a **failure**

4. a **log** cabin; **logged** 30,000 air miles in April; a trip **log**; **log** statistics on the computer

5. a **drone** pilot; that politician's voice was a relentless **drone**; the **drone** of an airplane; her voice **droned** on

Task 3 Word family

Directions: Based on the words in the left column, find out the meaning of each word in the right column.

1. driverless	meaningless; friendless; homeless; sleepless; lifeless; childless; purposeless
2. disclosure	disapproval; disadvantage; disappointment; disbelief; discount; disfavor
3. mismatch	misunderstand; misguide; mispronounce; misinterpret; mishear; miswrite
4. counterbalance	counter-clockwise; counterstrike; counteract; counterproductive; counter-terrorism
5. well-funded	well-said; well-done; well-written; well-played; well-dressed

III Speaking

Team discussion and peer feedback are pivotal to learning new ideas, inspiring deeper thinking, sharing best practices, brainstorming creative solutions as well as strengthening effective communication and listening skills.

Task 1

Directions: Give a 2-minute talk on one of the following topics.

1. What lessons can you draw from Edward Snowden's NSA disclosures?

2. What are some of the Internet regulation problems that've bothered you the most? What might be the possible measures to tackle those problems?

3. How do you think we can build a government structure to tackle the securing hazard at a systemic level?

Task 2 Structured discussions

Directions: Work in groups of 3 to 4 and review the reading material "Security and the Internet of Things". Each person highlights at least one significant idea and/or quote in the text.

And then within your group:

a. Identify a facilitator/time keeper.

b. The first person presents the idea or quote they highlighted and why they think it is important. (3 minutes)

c. Each person (1 minute) responds to what they heard from the presenter and the presenter does not talk and only takes notes.

d. The presenter is given one more minute to respond to what they heard or clarify their thinking. (1 minute)

e. Take turns to be the presenter.

Ⅳ Writing

Asking good questions to each other is an important communication skill for the success of you and your team in the future workplace. In order to ask effective questions that elicit thinking, you need to:

a. Use the approachable voice;

b. Employ plural forms;

c. Use tentative/exploratory language;

d. Embed positive presuppositions;

e. Design open-ended and probing questions.

Task 1

Directions: Work in pairs. First, work on your own and write down as many effective questions related to the text as possible, and then take turns to ask and answer each other's effective questions.

____
____
____
____
____
____

Task 2

Directions: Choose one of effective questions that you've designed or heard or are listed as below and write down your thoughts on it.

Approach	Possible Effective Questions
Past strategy	Have you overcome a situation where you were hacked? What did you do?
Hypothetical	If you were a hacker, what bugs or loopholes would make you easily hack into a computer system?
Fresh start	If you could erase or rewind everything and start over, how would you build a collective governance institution that deals with the security hazards posed by IoT?

_____
_____
_____
_____
_____
_____
_____
_____
_____

KEY

I Reading

Task 1

Logic Map of
"Security and the Internet of Things"

⦿ **What is changed by IoT:** We have computers with things attached to them.

(para. 1)

Examples:
- refrigerator—a computer that keeps things cold.
- oven —a computer that makes things hot.
- ATM—a computer with money inside.
- car—a computer with four wheels and an engine.

(para. 2)

⦿ **What is IoT:** (para. 3~6)

a. IoT is a computerized, networked and interconnected world that we live in.

(para. 3)

b. IoT is a world-size robot with three parts:

$$\left.\begin{array}{l}\text{the } \underline{\text{sensors}} \text{ —eyes of ears of the Internet} \\ \text{the "} \underline{\text{smarts}} \text{"—brain of the Internet} \\ \text{the } \underline{\text{actuators}} \text{ —hands and feet of the Internet}\end{array}\right\}$$ （para. 4～6）

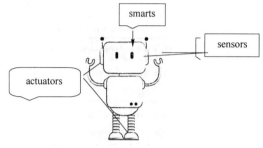

◉ **What is the security concern of IoT：**

a. IoT is <u>hackable</u> just like computers. （para. 7）

b. <u>Markets</u> alone can't solve our security problems as they are motivated by <u>profit</u> and <u>short-term goals</u> at the expense of society. （para. 8）

◉ **What are the general solutions：**

a. We should get the <u>policy</u> right. （para. 9）

b. As policy change means significant government regulation, a new <u>government regulatory agency</u> is needed. （para. 10）

◉ **What's wrong with the current Internet regulation：**

a. There is no government structure to tackle the security hazard at a <u>systemic</u> level. （para. 11）

b. Each agency has a <u>different</u> approach and <u>different</u> rules. （para. 12）

IoT needs to be viewed as a <u>single entity</u> and any solutions here need to be <u>holistic</u>.

（para. 13～14）

◉ **What should the government do：** （para. 15）

Government needs to

$$\left\{\begin{array}{l}\text{ensure companies } \underline{\text{follow good security practices}} \text{ and hold them } \underline{\text{liable}}. \\ \text{mandate strong } \underline{\text{personal data}} \text{ protections and limitations on } \underline{\text{data collection and use}}. \\ \text{legalize and fund } \underline{\text{responsible security research}}. \\ \text{ensure } \underline{\text{transparency}} \text{ and } \underline{\text{interoperability}}.\end{array}\right\}$$

◉ **A summary：**

IoT is less designed than created and we need to rebuild confidence in <u>our collective governance institutions</u>.

（para. 16）

Ⅱ Vocabulary

Task 1

1. in; to   2. to   3. through   4. between; and   5. out of   6. of; at   7. in

Task 2

1. 我穿一身**整洁**的海军蓝制服；**时髦**的晚宴；我的双眼被烟熏得**刺痛**；**行家的投资**；街头**智慧**；**读书好**（有时指死读书）

2. 把手**插进**口袋里；把包装**满**；吃 5 个汉堡**填饱**肚子；**标本**鸟；**填充**玩具；鼻子**堵住**了

3. 企业**倒闭**；引擎**故障**；婚姻**失败**；他**未能**通过考试；他认为自己是个**失败者**

4. **木屋**；四月**航行**了三万航空里程；旅行**日记**；把数据**录**在计算机上

5. 无人机**飞行员**；那个政客不停地**唠叨**；飞机的**嗡嗡声**；她**唠叨**个不停

Task 3

1. 没有意义地；没有朋友；无家可归的；失眠的、难以入睡的；无生命的、无生机的；无子女的；无目的的

2. 不赞同；劣势、缺点；失望；怀疑；打折扣、低估；不赞成、冷待

3. 误解；误导；发错音；曲解；听错；写错

4. 逆时针；反击；抵消、中和；事与愿违的；反恐怖主义

5. 说得好；做得好、全熟；写得好；（比赛）发挥好；穿着得体

Unit 14　The Internet of Things Is Revolutionizing Our Lives, but Standards Are a Must

scalability /ˌskeɪləˈbɪlɪtɪ/ *n.*
the ability of a business or system to grow larger

visionary /ˈvɪʒənərɪ/ *n.*
a person with unusual powers of foresight

burst /bɜːst/ *v.*
to break open or apart suddenly, or to make something do this

mainstream /ˈmeɪnˌstriːm/ *n.*
the way of life or set of beliefs accepted by most people

heating /ˈhiːtɪŋ/ *n.*
the system that keeps a building warm

elephant in the room an obvious truth that is being ignored or goes unaddressed

proprietary /prəˈpraɪətərɪ/ *adj.*
relating to owning something, or relating to or like an owner

complexity /kəmˈplɛksɪtɪ/ *n.*
the state of having many parts and being difficult to understand or find an answer to

associated /əˈsəʊsɪeɪtɪd/ *adj.*
connected

1　The beauty of what IoT promises is a seamless interoperability between things, **gateways**, applications and users that will enable scalability, cross things **added value applications**, services and **cost optimization**. The IoT, however, may stall unless providers collaborate to link up different services.

2　"The Internet of Things" was first coined by a British visionary called Kevin Ashton. Back in 1999, before the **dotcom bubble** started to burst, he first used it to describe how **Internet-connected devices** would change our lives.

3　Ashton forecasted a futuristic world of seamlessly connected devices that would save us both time and money. Fast forward 15 years and the idea has become mainstream. Households across Britain are comfortable with brands like Nest and Hive connecting their heating systems to the Internet. You can expect to see more and more connected technology in your home in the near future.

4　But there is an elephant in the room. The road to mass adoption has not been smooth. Like many booming areas of technology before it, the IOT revolution is plagued by a lack of industry standards. Or else, our various smart devices would not be able to "talk" to each other.

5　If, like many early adopters, you have packed your home with the latest in smart technology, you will find few of your devices can communicate with one another. Want to turn your lights on and off? You will need Phillips' own proprietary app. Heating on too low? Better open your Nest app.

6　Standardization is one of the most critical hurdles of the IOT evolution. Without global standards, the complexity of devices that need to connect and communicate with each other (with all the associated addressing, automation, quality of service, **interfaces, data repository** and associated **directory services**, etc.) will grow exponentially. The IoT promises billions of connected things which in turn require common standards in order to operate with an acceptable, manageable and scalable level of complexity.

7 Let's take a simple transaction: data is typically collected by sensors within IoT devices, transmitted via wireless and wired networks to data and application warehouses <u>virtualized</u> in the cloud, and aggregated for analysis to determine information such as **usage patterns** through **analytics** and **business intelligence applications**. Standardizing is not particularly important to solve interoperability, as this can always be achieved through **multiprotocol gateways**. Standardization across the IoT landscape is important because this reduces the gaps between protocols (and associated **security holes** and issues). It also reduces the overall cost of data, associated transport costs and the cost of <u>manufacturing</u> of individual components. This is because fewer standards enable more compatible components, which all leads to reduced cost of design, manufacturing and a reduced **time to market**. Standardization also <u>streamlines</u> the overall integration at an application level (aggregation of data, interoperability of data, reports and business processes) without being concerned with individual IT devices, unique protocols and non-standard **data formats**.

8 Critics point out that many of positive benefits of a smart home are <u>outweighed</u> by the inconvenience of having to open separate apps for every device. Major players in the space have built and protected their own ecosystem rather than sign up to **industry-wide standards** for the benefit of consumers and the industry. Some predict a repeat of the VHS v. s. Betamax and HD-DVD v. s. Blu-ray standards wars that harmed both consumers and manufacturers.

9 The battle for <u>supremacy</u> has started. Late last year, Apple announced **HomeKit**, a <u>centralized</u> control system tied to the **iPhone ecosystem**. Critics point out that Apple's system will likely not <u>interact</u> with Google's competing Nest ecosystem. Samsung have also joined in the race, creating further <u>chaos</u> for consumers.

10 This becomes more complicated when multiple devices communicate with each other, with various data aggregators, and with multiple command and control stations across different countries and associated jurisdictions and laws. In the world of **telemedicine**, which heavily depends upon <u>connectivity</u>, medical devices could be operated from within the patient's room as well as from a central location in the hospital, and possibly even from

---

virtualize /ˈvɜːtʃuəˌlaɪz/ v.
to change something that exists in a real form into a virtual version (= one that is created using a computer)

manufacturing /ˌmænjʊˈfæktʃərɪŋ/ n.
the business of producing goods in large numbers

streamline /ˈstriːmˌlaɪn/ v.
to shape something so that it can move as effectively and quickly as possible

outweigh /ˌaʊtˈweɪ/ v.
to be greater or more important than something else

supremacy /sʊˈprɛməsɪ/ n.
the leading or controlling position

centralized /ˈsɛntrəlaɪzd/ adj.
controlled by one main system or authority

interact /ˌɪntərˈækt/ v.
to communicate with or react to

chaos /ˈkeɪɒs/ n.
a state of total confusion with no order

connectivity /ˌkɒnɛkˈtɪvɪt/ n.
the ability of a computer, program, device, or system to connect with one or more others

privileged /ˈprɪvəlɪdʒ/ *adj.*
having or enjoying one or more privileges

specify /ˈspɛsɪfaɪ/ *v.*
to explain or describe something clearly and exactly

release /rɪˈliːs/ *v.*
to allow a substance to flow out from somewhere

all-out /ˌɔːlˈaʊt/ *adj.*
complete and with as much effort as possible

definition /ˌdɛfɪˈnɪʃən/ *n.*
a statement that explains the meaning of a word or phrase

fragmented /fræɡˈmɛntɪd/ *adj.*
consisting of several separate parts

domain /dəˈmeɪn/ *n.*
an area of interest or an area over which a person has control

a remote location on the other side of the planet. Standards are vital to allow these devices to operate with each other, with a physician's tablet in another country, and within the hospital control system. Additionally, since the information is privileged and private, regulatory standards are required to ensure the data remains secure, and to specify the ownership and under what conditions that it can be released to others.

11   But unlike the standards battles of the past, the IoT revolution may not need an all-out winner. Intamac's Mark Lee thinks the trend is instead on interoperability between different systems. "Services like Spotify Connect, which in turn can control Sonos **wireless speakers** as well as other **Bluetooth** and Apple Airplay speakers in your home offer a better solution for consumers and businesses alike." Collaborations like these are currently the exception rather than the rule.

12   Until now, a few large organizations have been responsible for the definition of the **M2M protocols** or have performed the needed research and development, and the **market presence** to enforce those standards. This has created a **multiplicity** of M2M **stacks** and ecosystems with very limited interoperability. The current M2M related standards landscape is highly fragmented. This can be seen across domains where only basic communication and networking standards are being used. There is a trend to increase coordination work done to define a common IoT stack by the different international standard organizations. But there is still a long road ahead of us.

13   There is no doubt the IOT will continue to become an important part of our digital lives. But without the major players working together, consumers may lose interest and this revolution could stop in its infancy. If common standards are achieved, Kevin Ashton's futuristic vision of our connected world could soon become a reality.

Word count: 993

Source: Wood A. The Internet of Things is revolutionizing our lives, but standards are a must [EB/OL]. (2015-03-31). http://www.theguardian.com/media-network/2015/mar/31/the-internet-of-things-is-revolutionising-our-lives-but-standards-are-a-must.

I Reading

A jigsaw activity is a group activity in which each member is dependent on the others for part of the information. In other words, in order to complete a task, group members must cooperate. A jigsaw provides a good way for members to learn new materials and also provides an opportunity for members to teach each other what they have learned. **Jigsaw reading** is a cooperative learning strategy that enables each student of a "home" group to specialize in one aspect of a topic.

Task Jigsaw reading

Directions: Read the passage for 15 minutes, with each group bearing a specific task in mind. Discuss within your group, and then share with the whole class.

Group 1: Find out benefits brought by IoT: description and example.

Group 2: Find out drawbacks of IoT due to a lack of industry standards: description and example.

Group 3: Find out the benefits of IoT Standardization.

Group 4: Find out why the situation becomes more complicated when multiple devices communicate with each other.

Group 5: Describe what the world will be like if standardization is put into place. You can combine your prior knowledge about the topic and collaborate to make an evidence-based hypothesis.

II Vocabulary

Key words and expressions

scalability	visionary	burst	mainstream
heating	elephant in the room	proprietary	complexity
associated	virtualize	manufacturing	streamline
outweigh	supremacy	centralized	interact
chaos	connectivity	privileged	specify
release	all-out	definition	fragmented
domain			

Terminology

gateway	n.	网关
added value applications	n.	增值应用
cost optimization	n.	成本优化
dotcom bubble	n.	网络泡沫
Internet-connected devices	n.	互联网连接设备
interface	n.	界面；接口
data repository	n.	数据储存库
directory service	n.	目录服务；查号服务
usage pattern	n.	使用模式
analytics	n.	分析学；解析学
business intelligence applications	n.	商业智慧应用层
multiprotocol gateway	n.	多协议网关
security hole	n.	安全漏洞
time to market	n.	上市时间
data format	n.	数据格式
industry-wide standards	n.	行业标准
HomeKit	n.	智能家居
iPhone ecosystem	n.	iPhone 生态系统
telemedicine	n.	远程医学
wireless speaker	n.	无线扬声器
Bluetooth	n.	蓝牙技术
M2M protocol	n.	M2M 协议
market presence	n.	市场形象
multiplicity	n.	多重性
stack	n.	堆

Task 1 Word bank

Directions: Please fill in the blanks with the words listed below. You may need to change the form when necessary.

fragmented	burst	specify	all-out	domain
outweigh	heating	release	streamline	interact

1. A dam _____ and flooded their villages.
2. Virtualization of protocol and pattern implies conversion between different communication protocols and _____ patterns.
3. They're making efforts to _____ their normally cumbersome bureaucracy.
4. Make sure that all gas fires and central _____ boilers are serviced annually.
5. The business is looking for ways to _____ production.
6. Huawei P30 Series smartphones have been _____ on April 11, 2019.
7. Do you think the disadvantages of this change _____ its advantages?

8. Scientists are tracking the genetic _____ of COVID-19.

Task 2 Words in context

Using context clues to understand the meaning of words is an effective way to save your time when reading. It is also a great way to expand vocabulary.

Directions: Try to infer the meaning of the bolded word in the sentence.

1. The IoT, however, may **stall** unless providers collaborate to link up different services. (   )

   A. advance　　　　B. stop　　　　C. establish

2. "The Internet of Things" was first **coined** by a British visionary called Kevin Ashton. (   )

   A. copy　　　　B. purchase　　　　C. invent

3. Ashton forecasted a **futuristic** world of seamlessly connected devices that would save us both time and money. (   )

   A. modern　　　　B. beautiful　　　　C. future

4. Like many booming areas of technology before it, the IoT revolution is **plagued** by a lack of industry standards. (   )

   A. play　　　　B. annoy　　　　C. popularize

5. Until now, a few large organizations have been responsible for the definition of the M2M protocols or have performed the needed research and development, and the market presence to **enforce** those standards. (   )

   A. obey　　　　B. develop　　　　C. violate

6. But without the major players working together, consumers may lose interest and this revolution could stop in its **infancy**. (   )

   A. golden age　　　　B. adolescence　　　　C. early stage

Task 3 Word Family

*Directions: Look at the description from paragraph 6: "The IoT promises billions of connected things which in turn require common standards in order to operate with an **acceptable, manageable and scalable** level of complexity."*

Brainstorm other -able/ible words that can be used to describe technology. Try to define each word in English and make a sentence for each. An example has been given to you.

Word	English meaning	Sentence
portable	easily or conveniently transported	It is advisable to equip some teachers with a portable wireless speaker

Ⅲ Speaking

Task 1 Conversation

Directions: *Have a conversation with your partner. Talk about the connecting systems in your homes and anticipate the future.*

Reference: paragraph 3: "Households across Britain are comfortable with brands like Nest and Hive connecting their heating systems to the Internet. You can expect to see more and more connected technology in your home in the near future."

Task 2 Debating: four corners activity

Directions: *Discuss a statement written in the middle of the chart: decide your opinion and then move to a corner of the room to form groups. Each group prepares to present your opinion. After the presentations you can change sides. Say why you have changed your mind.*

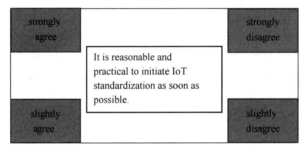

Here are a few ways to argue logically:

- Give examples;
- Compare and contrast your argument with the opposition's argument;
- Point out cause-and-effect relationships;
- Argue by definition.

Support your arguments with evidence:

- Personal experiences;
- The experiences (testimonials) of others; experts in the field;
- Facts gathered from research;
- Reliable statistics;
- Charts, graphs and pictures;
- Hypothetical examples.

Ⅳ Writing

Write a Persuasive Letter

1. What is a persuasive letter?

Do you know the reason why the persuasive letter is written? The persuasive letter is written in order to influence the action or thought of a reader. This can be done by providing appropriate reasons or by showing some actions. While writing a persuasive letter, you can support your opinion by giving certain examples. In such a letter, it is essential for you to state your opinion clearly. You can write such a letter to any organization or any individual. You can write it to any type of organization, such as the school, the bank, the NGO, the college, the municipality, etc. You can write it to any individual like: the CEO, the director, the government officer, etc.

The main motive of writing a persuasive letter is to get your work done. The persuasion can be related to anything like sale, complaint, petition, etc. It is essential for you to write the letter in a convincing manner. Some of the letters like a cover letter, request letter, complaint letter etc. all belong to the type of persuasive letter. In all such letters, the maximum amount of persuasion is involved from the side of the sender.

2. How to write a persuasive letter?

If you want the reader to agree with you, it is essential to write a persuasive letter in a convincing manner. Here are the steps that you need to take into account while writing a persuasive letter.

- First step

What is the topic of persuasion?

It is not an easy task to persuade anyone. Before you start your letter, it is very important for you to know complete details about the subject matter. What do you want and why? Give answers to these questions in your persuasive letter. If you know the topic properly in depth, then you can make the best arguments.

For instance, you want to convince your parents to buy a dog, then it is very important for you to focus on the main topic.

- Second step

Do you know your reader well?

Once you know the purpose of writing a persuasive letter, your next step will be to know your reader. Try to find the reader's belief, nature, weak points, and strong points, because this can make it easy for you to draft a letter in a convincing style. Figure out the concerns of the reader and try to offer a solution, accordingly.

For instance, if you think your parents are against buying a pet, then you can persuade them by stating some positive aspects related to the pet.

- Third step

Is your argument relevant?

If you think that your argument is relevant and correct, then put your best foot forward. Try to justify your arguments by giving proper reasoning. It is important for you to support your arguments with facts.

For instance, your parents might think that you are not capable or responsible enough to handle a pet. In such scenario, you should provide them with all the details related to the pet, such as food, habits, and schedule. Show the knowledge that you have about the pet because this can support your argument.

- Fourth step

How important is it to make the final appeal?

Don't just end your letter in a rude or impolite tone. Make final appeal before ending your letter in a convincing tone, so that the reader can give a second thought to your need or argument.

3. Persuasive letter writing tips

Before starting with a persuasive letter, there are a few things you need to keep in mind. Here are a few tips that can help you to draft an appropriate persuasive letter.

- Are you really sure about the persuasion? Do you think that your argument is valid? Getting answers to these questions can make it easy for you to write a letter with a right attitude.
- Conduct some research before starting with your persuasive letter.
- Outline the main things that you want to add in your letter. Organize the entire matter carefully, so that the reader can find it easy to connect to it.
- Collect the relevant information that supports the persuasion because this can help you to convince the reader.
- It is important for you to be clear with your needs because this can easily justify the conviction.
- It is also important for you to know the personality of the recipient because this can help you to write the letter in a right style.
- If you want the response of recipient to be in your favor, then it is important for you to understand the subject matter in depth.
- You should try to collect information from different sources because this can help you to have enough knowledge about the subject matter and this knowledge can help you to write a letter in the correct manner.
- When you are trying to prove your argument, you should try to be as objective as possible. As a result, it is important to figure out some pros and cons.
- List out a few pros and cons so that you can draft your persuasive letter on the basis of them.
- In order to get the basic idea of drafting such a letter, you can go through different formats and templates. Going through some persuasive letters can give you some ideas about the writing skills.

4. Persuasive letter elements and template

- Your name
- Your address

- Recipient's organization
- Recipient's name
- Recipient's address
- Subject line
- Salutation
- First paragraph (introductory lines)
- Second paragraph—(persuasive analysis)
- Third paragraph—(justification of your persuasion)
- Letter closing
- Your signature
- Your first name and last name
- Your designation
- Enclosure
- Logo of your organization (optional)

Date

    Sender's Name
    Address
    Contact Number
    Email

    To
    Addressee's Name
    Address
    Subject: Persuasive Letter for (mention nature of persuasion)

    Distinguished Sir/Madam,
    First Paragraph
    (introductory lines)
    Second Paragraph
    (persuasive analysis)
    Third Paragraph
    (justification of your persuasion)

    Letter closing
    Your signature
    Your full name
    Your designation
    Enclosure
    Logo of your organization (optional)

5. Sentence structures that you can use

- **Beginning**：

I'm writing to tell you about... We have had...

I have been thinking about the question you wrote in the letter. In my opinion, you should...

I'm writing to you to know what is going on with...

It is strongly recommended that...

As to the problem you mentioned in the last letter, here are some suggestions.

My suggestions are as follows.

It is my pleasure to offer you some advice on...

Since you have asked me for advice regarding... I will try to make some constructive suggestions.

When it comes to such a situation, I would like to suggest that you should...

If I were you, I would...

- **Ending**：

I feel that it would be of great benefit if...

I would appreciate it if you could follow my suggestions.

I would be very pleased if you could take my advice.

It would be an honor for me if you could take my proposal into account.

I hope you can benefit a great deal from my suggestions.

In the end, I hope you will find my suggestions useful.

It would be very nice if you could take my suggestions into account.

Task   Writing a persuasive letter

Directions: *Suppose you are an IT manager working for an international hospital. Your hospital has employed several IoT systems from different providers to satisfy different needs. For each system, you need a team of technicians to support and operate, which is costly. Therefore, you have been plagued with the lack of IoT standardization and desired it very much. Please write a Persuasive Letter to executives of some big technology companies or organizations calling for standardization.*

_____
_____
_____
_____
_____
_____
_____

KEY

II Vocabulary

Task 1

1. burst 2. interaction 3. revolutionize 4. heating 5. streamline 6. released 7. outweigh 8. evolution

Task 2

1. B 2. C 3. A 4. B 5. A 6. C

Task 3

portable
accessible
affordable
compatible
convertible
fashionable
...

Unit 15 What the Internet of Things Means for Consumer Privacy

unfold /ʌnˈfəʊld/ v.
to open or spread out something that has been folded
steady /ˈstɛdɪ/ adj.
happening in a smooth, gradual, and regular way, not suddenly or unexpectedly
aware /əˈwɛə/ adj.
knowing that something exists, or having knowledge or experience of a particular thing
adept /əˈdɛpt/ adj.
having a natural ability to do something that needs skill
expand /ɪkˈspænd/ v.
to increase in size, number, or importance, or to make something increase in this way
ubiquity /juːˈbɪkwətɪ/ n.
the fact that something or someone seems to be everywhere
invisibility /ɪnˌvɪzəˈbɪlətɪ/ n.
the fact of being impossible to see, or of not being noticed
transmit /trænzˈmɪt/ v.
to broadcast something, or to send out or carry signals using radio, television, etc
documentation /ˌdɒkjʊmɛnˈteɪʃən/ n.
the instructions for using a computer device or program
transfer /trænsˈfɜː/ n.
moving someone or something from one place, vehicle, person, or group to another
tricky /ˈtrɪkɪ/ adj.
difficult to deal with and needs careful attention or skill

come into force go into effect

1 As the digital era has <u>unfolded</u>, consumers have become <u>steadily</u> more <u>aware</u> that businesses would make use of their personal information. Besides, many consumers have become <u>adept</u> at exercising control over how their data are used. However, the IoT—the rapidly <u>expanding</u> network of devices, physical objects, services and applications that communicate over the Internet—poses a new set of privacy challenges, as it changes the relationship between individuals and their personal data.

2 The biggest challenges are <u>ubiquity</u> and <u>invisibility</u>: connected devices number in the billions today, and they <u>transmit</u> data without device owners knowing when or how that happens. The **data custody chains**, or <u>documentation</u> recording the <u>transfer</u> of data to different parties, are also complex. "The IoT combines the technologies of multiple providers, which makes the tracking of collected personal data extremely difficult, if not impossible, in most cases," says Giulio Coraggio, partner and head of global IoT and gaming at **DLA Piper**, a law firm. The same issues make the privacy challenges of the IoT difficult for government and industry to address. It is the cross-sector and crossborder **interlinkages** that make the IoT such a <u>tricky</u> area for stakeholders to grasp and address.

3 The **EIU** has conducted a survey of over 1,600 consumers in eight countries. The report draws on the analysis of the results and discusses how industry and government can help to build consumer trust in the age of IoT.

Consumer demands for privacy rights

4 Judging by the survey, many consumers globally want the types of data privacy rights that EU citizens will enjoy when **GDPR** <u>comes into force</u>. When asked to cite the most important rights regarding third-party use of their personal information, the majority of consumers (57%) most frequently cite **the right to erasure** of their information (also known as "the right to be

forgotten"), followed by **the right to object** to the use of their personal data, and to be informed in a clear way how the data are being used.

5   According to Mr. Coraggio: "GDPR grants individuals much stronger tools, such as the ability to launch class action claims against companies that exploit consumers' personal data." Other experts also agree that IoT-relevant privacy prescriptions are well enshrined in GDPR, and point out that they apply to all companies that process EU citizens' data. Many US and Asian companies, then, whether or not they have a physical EU presence, will need to abide by GDPR.

6   On a regional level, the right to erasure is most frequently cited by European consumers (61%), and it remains at the top of consumers' lists in **APAC** (56%) and the US (50%). Demand for the right to object is weaker, however, in APAC (39%) than in Europe or the US (50% in each). Notwithstanding these few differences, consumers in all three regions place the greatest weight on those rights that address fundamental issues of transparency and control.

What industry and government can do

7   If proactive consumer management of their personal data proves to be impractical in the IoT context, confidence building measures in the IoT's integrity are well within the ability of industry and government. Improved transparency is realistic, and is a good starting point for building trust. If consumers are adequately informed of how their personal data are processed, it should be possible to build their confidence in the IoT. Efforts such as posting simple notices or electronic alerts that devices are autonomously collecting data are small steps towards building transparency. They could help device manufacturers and service providers improve customer relationships and enhance their brand.

8   At the very least, some consumers would like such companies to publicly commit to maintaining consumer privacy. Cross-industry standards on delivering such transparency and other privacy protections would also help to earn consumer trust.

9   Nearly a third of respondents say that a rigorously upheld industry-led commitment to privacy protection would be

---

grant /grɑːnt/ v.
to give or allow someone something, usually in an official way

claim /kleɪm/ n.
a statement that something is true or is a fact, although other people might not believe it

prescription /prɪˈskrɪpʃən/ n.
rules or a situation that will have a particular effect

be enshrined in if a political or social right is enshrined in something, it is protected by being included in it

abide by to accept or obey an agreement, decision, or rule

notwithstanding /ˌnɒtwɪθˈstændɪŋ/ adv.
despite the fact or thing mentioned

impractical /ɪmˈpræktɪkəl/ adj.
not sensible or realistic, and does not work well in practice

adequate /ˈædɪkwɪt/ adj.
enough or satisfactory for a particular purpose

alert /əˈlɜːt/ n.
a warning to people to be prepared

autonomous /ɔːˈtɒnəməs/ adj.
having the power to make decisions

commit /kəˈmɪt/ v.
to promise

rigorous /ˈrɪɡərəs/ adj.
demanding strict attention to rules and procedures

effective, either from individual companies or as a collectively maintained **"code of conduct"** across industries. A similar share of respondents demand that industry collaborate with governments to develop privacy standards and ensure their rigorous enforcement.

10 Enforcement of GDPR rules is the job of each EU member's supervisory authority, which in most cases is its data protection agency or regulator. The penalties it can apply if a company is found to be in breach are stiff: up to 4% of annual global turnover or €20m (whichever is greater).

11 According to some experts, monitoring and enforcement will not be effective unless countries establish oversight bodies that have responsibility for all aspects of digital consumer protection. Such bodies exist in many countries such as the UK, but few operate with the scope necessary to address the full range of challenges posed by digital technologies.

12 Voluntary standards and guidelines agreed by multiple stakeholders would do much to build trust in the IoT but these typically require leadership from institutions such as the **UN**, **OECD** or **International Organization for Standardization** and can often take several years to complete. The international UN guidelines on consumer protection include some digital elements, particularly related to e-commerce, which can act as a starting point for future development of IoT standards.

Conclusion

13 People hold different opinions on whether IoT-specific privacy legislation is needed, but experts and consumers appear to agree that GDPR's provisions are a good starting point for countries looking to build concrete privacy safeguards relevant to the IoT. There is also a broad consensus that, along with IoT-related bodies of privacy standards and guidelines, close monitoring is needed to ensure adherence by device manufacturers and service providers. These messages come through clearly from the consumers in our survey and the experts we interviewed.

14 It is also apparent that more education and knowledge-building efforts are needed by all stakeholders involved, perhaps even before standards are developed. NGOs and a few government bodies have led the way in this effort, but manufacturers and service providers in the IoT **value chain** need to

---

collective /kəˈlɛktɪv/ *adj.*
of or shared by every member of a group of people

supervisory /ˌsuːpəˈvaɪzərɪ/ *adj.*
of or limited to or involving supervision

authority /ɔːˈθɒrɪtɪ/ *n.*
a group of people with official responsibility

regulator /ˈrɛɡjʊˌleɪtə/ *n.*
an official responsible for supervision of a particular activity

penalty /ˈpɛnəltɪ/ *n.*
a punishment for doing something against a law

breach /briːtʃ/ *n.*
the act of breaking a law, promise

stiff /stɪf/ *adj.*
firm or hard

turnover /ˈtɜːnəʊvə/ *n.*
the amount of business that a company does

monitor /ˈmɒnɪtə/ *v.*
to watch a situation carefully

oversight /ˈəʊvəˌsaɪt/ *n.*
the state of being in charge of sb/sth

voluntary /ˈvɒləntərɪ/ *adj.*
done, made, or given willingly, without being forced or paid to do it

guideline /ˈɡaɪdˌlaɪn/ *n.*
information intended to advise people on how something should be done or what something should be

concrete /ˈkɒnkriːt/ *adj.*
clear and certain, or real and existing in a form that can be seen or felt

safeguard /ˈseɪfɡɑːd/ *n.*
something that is designed to protect people from harm, risk or danger

consensus /kənˈsɛnsəs/ *n.*
a generally accepted opinion or decision among a group of people

adherence /ədˈhɪərəns/ *n.*
the fact of someone behaving exactly according to rules, beliefs, etc.

join in as well. Many have been vocal in discussions on IoT security, but less so when it comes to educating consumers about IoT privacy issues.

15 Multi-stakeholder agreement of IoT privacy standards is likely to take time, but educational initiatives targeted at both consumers and businesses should be widened, in terms of geography and sectors. Meanwhile, confidence-building measures, such as public company commitments to maintaining privacy or the posting of alerts that devices are collecting data, can be taken by businesses now. The need for such measures is urgent, as the IoT and other **data-crunching technologies** are moving ahead at great speed.

Word count: 1,109

Source: Adapted from Economist Intelligence Unit. What the Internet of Things means for consumer privacy [R]. London: EIU, 2018.

vocal /ˈvəʊkəl/ adj.
relating to or produced by the voice, either in singing or speaking

initiative /ɪˈnɪʃɪətɪv/ n.
a new plan or process to achieve something or solve a problem

I Reading

Task 1 True or false

Directions: Please write T or F for each statement based on the article.

____ 1. The IoT changes the way that individuals collect personal data.

____ 2. According to Giulio Coraggio, it is because the IoT integrates the technologies of different companies that make it hard to follow and document individual data in most situations.

____ 3. According to some experts, even if some US and Asian companies do not have branches within the EU, as long as they process EU citizens' data, they are obliged to observe GDPR.

____ 4. It is impractical to make consumers vigorously manage their personal data.

____ 5. Most consumers require that there should be collaboration between companies and governments for standards development and enforcement.

Task 2 Reading for details

Directions: Please answer the following questions.

1. Why does the IoT pose privacy challenges?

2. What make the privacy challenges of the IoT difficult for government and industry to address?

3. Why do many consumers globally want the types of data privacy rights that EU citizens will enjoy when GDPR comes into force?

_____

_____

4. In terms of the demand of consumers, what are the similarities among consumers in Europe, America and APAC?

_____

_____

5. Can you list three steps that consumers would like to see to improve privacy?

_____

_____

6. According to the article, what is the precondition of making monitoring and enforcement effective?

_____

_____

7. On which aspects do experts and consumers have consensus?

_____

_____

8. Stakeholders have participated in lots of discussions on IoT security, but in which aspect do they need to scale up efforts?

_____

_____

Task 3

Directions: Read paragraph 7~9 and fill in the blanks.

	Measures	Benefits
1	improved _____	earn _____
2	companies publicly committing to _____	
3	cross-industry standards on _____	
4	a rigorously upheld _____ commitment to privacy protection	
5	_____ to develop privacy standards	

Task 4

Directions: The following sentences (extracted from the concluding part) are in the wrong order. Number the sentences to create a logical sequence. You can make use of lexical clues to help you. (For example, before you start, you may look at the final sentence of the previous paragraph to help you decide the sequence.) *Then check your rearrangement against the original order in the text.*

1. There is also a broad consensus that, along with IoT-related bodies of privacy standards and guidelines, close monitoring is needed to ensure adherence by device manufacturers and service providers. ☐

2. People hold different opinions on whether IoT-specific privacy legislation is needed, but experts and consumers appear to agree that GDPR's provisions are a good starting point for countries looking to build concrete privacy safeguards relevant to the IoT. ☐

3. NGOs and a few government bodies have led the way in this effort, but manufacturers and service providers in the IoT value chain need to join in as well. Many have been vocal in discussions on IoT security, but less so when it comes to educating consumers about IoT privacy issues. ☐

4. These messages come through clearly from the consumers in our survey and the experts we interviewed. It is also apparent that more education and knowledge-building efforts are needed by all stakeholders involved, perhaps even before standards are developed. ☐

Ⅱ Vocabulary

Key words and expressions

unfold	steady	aware	adept
expand	ubiquity	invisibility	transmit
documentation	transfer	tricky	come into force
grant	claim	prescription	be enshrined in
abide by	notwithstanding	impractical	adequate
alert	autonomous	commit	rigorous
collective	supervisory	authority	regulator
penalty	breach	stiff	turnover
monitor	oversight	voluntary	guideline
concrete	safeguard	consensus	adherence
vocal	initiative		

Terminology

data custody chain	n.	数据保管链
DLA Piper	n.	欧华律师事务所
interlinkage	n.	互连；链接
EIU(Economist Intelligence Unit)	n.	经济学人智库(经济学人集团旗下智囊机构)
GDPR (General Data Protection Regulation)	n.	通用数据保护条例
the right to erasure	n.	删除的权利
the right to object	n.	反对的权利
APAC (Asia-Pacific)	n.	亚太地区
code of conduct	n.	行为准则；规范
UN (United Nation)	n.	联合国
OECD (Organization for Economic Co-operation and Development)	n.	经济合作与发展组织
International Organization for Standardization	n.	国际标准化组织
value chain	n.	价值链
data-crunching technologies	n.	数据处理技术

Task 1 Word bank

Directions: Please fill in the blanks with the words listed below. You may need to change the form when necessary.

unfold	adequate	breach	supervisory	initiative	consensus
transfer	prescription	ubiquity	rigorous	turnover	notwithstanding
voluntary	abide	safeguard	stakeholder	adept	transmit

1. There are no financial institutions outside the _____ and regulatory system here.

2. The congressman was accused of a _____ of secrecy rules. The interests of minorities will have to be _____ under a new constitution.

3. We grow our organization based on interactions that promote mutual trust and respect with our _____ and partners.

4. Second, good deed belongs to _____ work; personal interests should not be the purpose of helping others.

5. China and the EU have many _____ on major international and regional issues.

6. If you join the club, you have to _____ by its rules.

7. The speed with which digital cameras can take, process and _____ an image is phenomenal.

8. The store greatly reduced the prices to make a quick _____.

9. Keynes had seen the future more clearly than most but when it came to how specific events would _____, he simply did not know.

Task 2 Word family

Most English words are formed in this way: root＋affixes, as is demonstrated by the following example: pre(前缀,之前的)＋dict(词根,说)＝predict(在……之前说;预言). We have more words with the same root: malediction, benediction, contradict, vindicate and so on. A good master of roots and affixes may multiply your vocabulary. You can build a mental picture of spider net by association.

Directions: Brainstorm words of the same root or affixes for some key words from the article and try to comprehend the meaning based on word formation. Pay special attention to the bolded part.

Unfold
un＋ _____

Transmit, **trans**fer
trans＋ _____

Exploit
ex＋ _____

Prescription
pre＋ _____

Impractical
im＋ _____

Collectively, **con**sensus, **con**crete
co＋ _____

Oversight
over＋ _____

Ⅲ Speaking

Task 1 Interview

Directions: According to the article, the EU has been a leader in the IoT standards. Suppose you are a TV host and you are going to interview an expert from EIU who have participated in standard development in the Dialogue program. Have a role-play with your partner, with one acting as the host and the other the expert. You may consult relevant document before the dialogue.

You may begin like this:

Host: Good morning, ladies and gentlemen, welcome to the *Dialogue*.
Today, we have Dr. Turk as our guest. Turk is…. Hello! Dr. Turk!
Turk: Hello! It's really nice to meet you.
Host: It is generally held that…. So…
Turk: …

Task 2 Public hearing

Directions: From the article we can learn that different stakeholders are involved in the IoT privacy standards. Suppose the government is going to hold a public hearing soliciting voices from different parties. And the main attendants include government officials, IoT companies and consumers. Accordingly, the class can be divided into the three groups to voice out demands and suggestions from different perspectives. One student may chair the public hearing; each group may identify one to guide discussion and another one to deliver group reports.

Topic: What can we do to better protect consumer privacy along with IoT development? You need to discuss: Benefits of the standards and what to do.

	Consumer	Government	Company
Benefit			
Difficulty			
What to do			

Expressions for reference: subsidized loan, facilitation of paperwork, business-starting consultation, tax deduction, guidance and direction, training, service platform

Ⅳ Writing

Creating a Research Agenda

1. What is a research agenda?

A document which contains information which is necessary for the planning and the execution of a study of research is known as research agenda. This is a document which can help researchers identify their tasks which need to be prioritized according to the phase of the research they are presently at. Besides, it also allows them to follow their particular research pattern. Moreover, it will be helpful for them to achieve the accurate results that they have been looking for. The research agenda becomes the guiding force for researchers to achieve their objectives eventually.

2. While drafting research agendas, you must provide:

- the kind of study that the researcher is required to undertake;
- the information useful for the researcher to accurately prioritize the main items of the research;
- any variables that are needed during the research;
- the processes of the research, the methods for collecting data, the community where the research will be applied along with the information analysis which will be used;
- an organized details within the document to ensure that an effective research agenda is provided to each individual within the team;

- any tips specifically related to the institution you are representing or information which you may have collected from experience;
- a certain degree of flexibility in terms of rules.

Research Agenda

Overview

The Challenge

Research Objectives

Activity	Objective	Potential Outcome
Task 1		
Task 2		
Task 3		

References

Task 1　Comment on measures

Directions: *Summarize some measures proposed in the article and write your comment for each from the perspective of consumers.*

_____
_____
_____
_____
_____
_____
_____
_____

Task 2　Research agenda

Directions: *The local government plans to develop a set of IoT privacy standards with a group of experts according to local conditions. You are one of the experts invited by the government and you are supposed to write a simple research agenda for the project.*

KEY

Ⅰ Reading

Task 1

1. F 2. T 3. T 4. F 5. F

Task 2

1. As it changes the relationship between individuals and their personal data.

2. The IoT combines the technologies of multiple providers, which makes the tracking of collected personal data extremely difficult, if not impossible, in most cases. The same issues make the privacy challenges of the IoT difficult for government and industry to address.

3. GDPR grants individuals much stronger tools, such as the ability to launch class action claims against companies that exploit consumers' personal data.

4. The right to erasure is most frequently cited; consumers in all three regions place the greatest weight on those rights that address fundamental issues of transparency and control.

5. Consumers would like such companies to publicly commit to maintaining consumer privacy; cross-industry standards on delivering such transparency and other privacy protections would also help to earn consumer trust; a rigorously upheld industry-led commitment to privacy protection would be effective.

6. Countries establish oversight bodies that have responsibility for all aspects of digital consumer protection.

7. GDPR's provisions are a good starting point for countries looking to build concrete privacy safeguards relevant to the IoT. Along with IoT-related bodies of privacy standards and guidelines, close monitoring is needed to ensure adherence by device manufacturers and service providers.

8. Educate consumers about IoT privacy issues.

Task 3

	Measures	Benefits
1	improved <u>transparency</u>	<u>earn consumer trust</u>
2	companies publicly committing to <u>maintaining consumer privacy</u>	
3	cross-industry standards on <u>delivering such transparency and other privacy protections</u>	
4	a rigorously upheld <u>industry-led</u> commitment to privacy protection	
5	<u>industry collaborate with governments</u> to develop privacy standards	

Task 4

2134

Ⅱ Vocabulary

Task 1

1. supervisory 2. breach, safeguarded 3. stakeholders 4. voluntary
5. consensuses 6. abide 7. transmit 8. turnover 9. unfold

Word List

abide by		遵守	U15
accelerate	v.	加速	U10
access	n.	进入,使用权	U4
acclimate	v.	使适应	U12
accuracy	n.	精度	U9
acronym	n.	首字母缩略词	U4
actuator	n.	执行机构	U5
acute	adj.	急性的	U7
adapt to		使适应	U11
address	v.	应对	U12
adept	adj.	熟练的	U15
adequate	adj.	充分的	U15
adherence	n.	遵守	U15
affix	v.	黏上	U8
aggregate	v.	总计	U9
aggregation	n.	聚集	U7
alert	n.	警示	U15
algorithm	n.	算法	U7
alleviate	v.	缓解	U9
alliance	n.	联盟	U4
all-out	adj.	完全的	U14
allude to sb/sth		暗指	U1
ambient	adj.	周围的	U5
analog	adj.	模拟的	U7
analogous	adj.	类似的	U2
analytics	n.	分析学	U9
anomaly	n.	异常	U7
appliance	n.	(家用)电器	U4
appropriate	adj.	合适的	U7
apt	adj.	合适的	U3

arbitrary	adj.	任意的	U11
around the clock		昼夜不停地	U7
array	n.	一大批，一大群	U5
arrhythmia	n.	心律失常	U7
asset	n.	资产,有用的东西	U4
associated	adj.	有关联的	U14
at hand		即将来临	U10
attain	v.	（经过努力）获得	U1
authority	n.	权威	U15
automated	adj.	自动化的	U4
autonomic	adj.	自主的	U4
autonomous	adj.	自动的	U15
aware	adj.	意识到的	U15
backbone	n.	主干	U3
base	n.	基础	U12
be associated with		和……联系在一起	U5
be credited to		归功于	U4
be enshrined in		纳入	U15
be grounded in/on sth		基于……	U1
bewilder	v.	使迷惑	U5
bewildering	adj.	使人迷惑的	U5
biometric	adj.	生物计量的	U8
blackout	n.	停电	U6
blade	n.	刀片	U2
bond	n.	纽带	U3
bottleneck	n.	瓶颈	U3
branding	n.	品牌推广	U12
breach	v.	违背	U15
breed	v.	滋生,导致	U11
burst	v.	破裂	U14
capture	n.	把……输入计算机；采集	U2
capture	n.	捕捉,捕获	U11
cardiac	n.	强心剂	U7
cardiovascular	adj.	心血管的	U7
catchy	adj.	引人注意的,容易记住的	U1
category	n.	类别	U10
centralized	adj.	集中的	U14
chaos	n.	混乱	U14
chronic	adj.	慢性的	U7
claim	n.	要求	U15

classify	v.	分类	U3
clinician	n.	临床医生	U7
clustering	n.	一群	U11
collaborate	v.	合作	U3
collective	adj.	共同的	U15
collocate	v.	并列	U5
combat	v.	反对；与……战斗	U4
come into force		生效	U15
commission	v.	委托	U10
commit	v.	承诺	U15
commonality	n.	共性	U12
commonplace	adj.	常见的	U7
commute	v.	（上班）通勤	U11
complementary	adj.	互补的，辅助性的	U1
complexity	n.	复杂性	U14
complication	n.	并发症	U7
component	n.	成分；组件	U4
computerized	adj.	计算机化的	U13
conceive	v.	构想	U10
concentration	n.	集中	U6
conceptualize	v.	使……概念化	U2
concrete	adj.	具体的	U15
confidentiality	n.	机密性	U4
confirm	v.	证实	U11
conformist	adj.	整齐划一的	U11
confusion	n.	困惑	U10
connectivity	n.	互联互通	U14
consensus	n.	共识	U15
consolidate	v.	巩固	U6
constrain	v.	限制	U3
converge	v.	汇集	U7
convergence	n.	（不同思想、群体或社会的）趋同，融合	U5
cooperative	adj.	合作的	U3
coordinate	v.	使协调，整合	U5
cosmetics	n.	化妆品	U10
counterbalancing	adj.	平衡的	U13
coverage	n.	覆盖	U3
cumulative	adj.	累积的	U6
customization	n.	定制	U12

daunt	v.	使害怕	U10
deactivate	v.	使无效	U11
deceptive	adj.	欺骗性的	U13
dedicated	adj.	专用的	U1
default	n.	默认值	U13
definition	n.	定义	U14
deliver	v.	交付	U10
demographic	adj.	人口学的	U6
demolish	v.	破坏,推翻	U13
dense	adj.	浓密的	U10
department	n.	部门	U12
dependability	n.	可靠性	U11
depiction	n.	描述	U9
deploy	v.	部署	U6
deployment	n.	部署	U9
desire	v.	想要	U11
detail	v.	详细列出	U12
detect	v.	探测,觉察	U11
detract from		损害,阻碍	U11
devastating	adj.	毁灭性的	U6
differentiate	v.	区分	U8
diffusion	n.	传播,扩散	U11
digitalis	n.	洋地黄	U7
dimension	n.	维度	U3
diminish	v.	减少	U1
disparate	adj.	迥然不同的	U4
dispense	v.	配(药)	U4
dispenser	n.	分配器	U8
disruptive	adj.	破坏性的	U9
disseminator	n.	传播者,传播媒介	U11
distribute	v.	分配	U3
distributed	adj.	分布式的	U13
diverse	adj.	多种多样的	U5
diversification	n.	多样化	U11
documentation	n.	文件记录	U15
domain	n.	主管领域	U14
domestic	adj.	国内	U9
dominant	adj.	处于支配地位的	U2
dormant	adj.	休眠的	U2
downtime	n.	(机器、设备的)停止运行期	U4

dramatic	adj.	突然的	U3
drone	n.	无人机	U6
dwell	v.	居住	U8
ecosystem	n.	生态系统	U4
efficient	adj.	高效的	U1
elephant in the room		显而易见却不愿触碰的事实	U14
eliminate	v.	根除	U7
emanate	v.	放射	U6
embed	v.	嵌入,内置	U1
embrace	v.	囊括	U1
emerge	v.	出现	U3
emission	n.	排放	U8
emit	v.	发射	U2
enable	v.	使能够	U10
encompass	v.	包含	U12
encryption	n.	加密	U4
enhance	v.	提高	U2
enormous	adj.	大量的	U10
enterprise	n.	企业	U4
ephemeral	adj.	短暂的	U11
estimate	v.	估计	U3
evacuate	v.	疏散	U6
evolution	n.	演化;发展	U1
expand	v.	扩展	U15
expedite	v.	加速	U6
expertise	n.	专业知识	U5
exploit	v.	利用	U10
explosion	n.	爆炸	U1
exponential	adj.	快速增长的	U2
exposure	n.	暴露	U9
externality	n.	外部性	U13
faucet	n.	自来水龙头	U8
fleet	n.	机群,车队	U4
forerunner	n.	先驱,前兆	U1
foreseeable	adj.	可预见的	U1
forethought	n.	先见,事先筹划	U13
format	n.	格式	U5
forward	v.	转发	U2
fragmentation	n.	破碎,分裂	U5
fragmented	adj.	零散的	U14

framework	n.	框架	U4
fraud	n.	欺骗,诈骗	U4
friction	n.	摩擦	U10
furnace	n.	熔炉	U13
gateway	n.	网关	U7
gauge	v.	测量	U8
geotarget	n.	地理定位	U6
glimpse	n.	初步的感受	U12
glucose	n.	葡萄糖	U7
grant	v.	给予	U15
granted	adv.	诚然	U12
grid	n.	输电网	U8
guideline	n.	准则	U15
hackable	adj.	容易被非法侵入的	U13
harmonize	v.	使和谐,使一致	U5
harvest	v.	收割;得到	U4
hazard	n.	危害	U9
heating	n.	供热	U14
heighten	v.	(使)加强	U9
holistic	adj.	整体的	U13
horizontally	adv.	横向地	U13
hub	n.	中心	U7
humidity	n.	湿度	U6
hypoxemia	n.	血氧不足	U7
identifier	n.	标识符	U5
imaginary	adj.	虚拟的	U11
immense	adj.	巨大的	U1
impede	v.	阻碍	U6
implant	v.	植入	U11
impractical	adj.	不切实际的	U15
in a… manner		以……的方式	U1
ins and outs	n.	复杂细节	U12
in retrospect		回顾	U1
in terms of		在……方面	U9
incoming	adj.	正来临的	U6
incorporate	adj.	合并的;一体化的	U4
incorporate	v.	合并	U7
incumbent	n.	在职者	U10
infiltrate	v.	渗入	U12
infrastructure	n.	基础设施	U8

ingest	v.	吸收	U10
inherent	adj.	内在的,自带的	U13
initiative	n.	倡议活动	U15
innovation	n.	改革,创新	U5
instill in		灌输	U11
integrate	v.	结合	U3
interact	v.	互动	U14
interaction	n.	(人与计算机或计算机与其他机器之间)交互作用,互动	U5
intermediary	n.	中间者	U1
interoperable	adj.	彼此协作的	U5
interpersonal	adj.	人与人之间的	U5
interpret	v.	解释	U1
intervention	n.	干预	U7
intoxication	n.	中毒	U7
intriguing	adj.	神秘的	U10
intrusion	n.	侵入;闯入	U4
inventory	n.	库存	U10
invisibility	n.	不可见(性)	U15
jurisdiction	n.	管辖范围	U13
launch	v.	启动	U10
leading-edge	adj.	先进的	U10
legislation	n.	立法	U5
lengthy	adj.	很长的	U11
leverage	v.	最大限度地利用	U6
liable	adj.	负法律责任的	U13
lightweight	adj.	轻量的	U9
line of sight		视线	U3
lobbyist	n.	说客	U12
log	v.	记录	U13
logistics	n.	物流	U10
loop	n.	环	U9
lousy	adj.	质量很差的	U13
low-lying	adj.	低洼的	U6
magnitude	n.	重大性	U10
mainstream	n.	主流	U14
maintenance	n.	维护	U3
mandate	v.	授权	U13
manual	adj.	手工的	U9
manufacturing	n.	制造	U14

measurement	n.	测量的结果	U4
mechanism	n.	方法；机制	U5
mediated	adj.	以……为介导的	U11
merchandise	v.	销售	U10
mesh	n.	网格	U4
metaphor	n.	象征	U1
minimize	v.	使减到最少	U3
mismatch	n.	不匹配，不协调	U13
modernize	v.	使现代化	U9
monetary	adj.	货币的	U8
monitor	v.	监督	U12
monitor	v.	监视	U15
monopolistic	adj.	垄断的，独占的	U13
muddle through		应付过去	U13
mutual	adj.	共同的	U5
myriad	n.	大量	U10
navigate	v.	确定方向	U10
navigation	n.	航行	U7
negligible	adj.	微不足道的	U1
norms	n.	准则	U11
notwithstanding	adv.	虽然，尽管	U15
obtain	v.	获得	U3
offer	n.	出价	U12
offline	adv.	离线，未连接的	U13
operational	adj.	操作的；运作的	U4
optimize	v.	优化	U1
outreach	n.	延伸	U6
outweigh	v.	超过	U14
overburdened	adj.	负担过重的	U6
overlap	n.	重叠	U5
oversight	n.	监察	U15
parallel	adj.	并行	U9
paranoia	n.	多疑，恐惧	U11
paranoid	adj.	多疑的	U12
passive	adj.	消极的	U1
patching	n.	补丁（计算机）	U13
patent	v.	授予专利；取得……的专利权	U4
penalty	n.	处罚	U15
perceive	v.	感知到	U1
personalization	n.	个性化	U12

personalize	v.	使个性化	U4
physical	adj.	实物的,有形的	U1
pillar	n.	柱子,支柱	U4
pinpoint	v.	指出	U6
pivotal	adj.	关键的	U2
portfolio	n.	档案	U6
postulate	v.	假定	U1
precision	n.	精度	U9
predict	v.	预报,预言	U4
preparedness	n.	已准备	U6
prescription	n.	规定	U15
prevalent	adj.	流行的,盛行的	U5
primate	n.	灵长目动物	U11
prioritize	v.	优先处理	U6
privileged	adj.	有特权的	U14
proactive	adj.	积极主动的	U4
process	v.	处理	U3
profitability	n.	盈利能力	U9
proliferation	n.	增殖	U9
propagate	v.	宣传	U1
proprietary	adj.	有所有权的	U14
protocol	n.	协议	U4
pursue	v.	追求	U10
quantity	n.	量,数量	U3
query	n.	问询	U10
radius	n.	半径	U6
random	adj.	随机的	U3
raw	adj.	原始的,未处理的	U2
realization	n.	实现	U3
real-time	adj.	实时的	U3
regulator	n.	调控者	U15
regulatory	adj.	管理的	U13
relay	v.	转播	U6
release	v.	释放	U14
relentless	adj.	不懈的	U10
reliable	adj.	可靠的	U3
requisite	n.	要素,要件	U5
respondent	n.	回应者	U10
restock	v.	补充	U4
retailer	n.	零售商	U10

retrieve	v.	检索	U8
revenue	n.	收益	U4
revolutionize	v.	彻底改变,革命	U1
rigorous	adj.	严格的	U15
roadblock	n.	路障	U10
robotics	n.	机器人学	U4
run low		不足,缺乏	U4
safeguard	n.	维护	U15
saturation	n.	饱和	U7
scalability	n.	可扩展性	U14
scalable	adj.	有扩展性的	U10
scenario	n.	场景	U9
seamless	adj.	无缝的,不停顿的	U2
segment	v.	分割	U4
self-expression	n.	自我表达	U11
self-fulfillment	n.	自我实现	U11
seminal	adj.	(在某一领域)有重大影响的	U2
sensor	n.	传感器	U4
setting	n.	环境	U4
shelter	v.	遮蔽	U3
shipment	n.	发货	U10
shorten	v.	缩短	U3
skillset	n.	技能组	U9
span	v.	跨越	U10
specification	n.	规范	U5
specify	v.	具体说明	U14
spectrum	n.	范围	U7
sphere	n.	领域	U9
stakeholder	n.	利益相关者	U8
stakes	n.	赌注,风险	U9
steady	adj.	稳定的	U15
stiff	adj.	难处理的	U15
stitch	v.	缝合	U10
stream	n.	流	U9
streamline	v.	简化	U14
subscribe	v.	订阅	U9
subset	n.	子集	U5
substantial	adj.	大量的	U1
subvert	v.	颠覆,推翻	U13
supervisory	adj.	管理的	U15

supremacy	n.	权威地位	U14
surrounding	adj.	周围的	U3
surveil	v.	监视	U6
surveillance	n.	监视	U11
synchronize	v.	同步	U9
tactic	n.	策略	U6
tag	n.	标签	U5
take stock of		观察,估量	U6
tap	v.	利用	U12
tap into		挖掘,开发	U4
telltale	adj.	报警的	U8
terminal	n.	终端	U5
the infirm	n.	年老体衰者	U11
thermostat	n.	恒温器	U13
tornado	n.	龙卷风	U6
totality	n.	全部,整体	U13
transfer	n.	移交	U15
transform	v.	改变	U10
transformational	adj.	转换	U9
transition	n.	过渡	U9
transmission	n.	传递;传送	U4
transmit	v.	传播	U15
transparency	n.	透明	U13
tricky	adj.	难对付的,狡猾的	U15
trigger	n.	引爆装置	U1
tsunami	n.	海啸	U6
turnover	n.	营业额	U15
ubiquitous	adj.	无所不在的	U1
ubiquity	n.	无处不在	U15
ultimately	adv.	最终地	U1
unauthorized	adj.	非法的;未被授权的	U4
underpin	v.	巩固	U8
underpinning	adj.	加强	U9
unfold	v.	展开,发展	U15
unverified	adj.	未经核对的,未经证实的	U4
upgrade	n.	升级	U12
utility	n.	公共设施	U4
utility pole		电线杆	U6
vend	v.	出售	U4
vending machine	n.	自动贩卖机	U12

vendor	n.	贩卖者	U10
vibrate	v.	颤动	U3
vicinity	n.	周围地区	U12
virtual	adj.	虚拟的	U1
virtualize	v.	虚拟化	U14
visibility	n.	能见度	U10
vision	n.	愿景	U1
visionary	n.	远见卓识者	U14
vital	adj.	至关重要的	U4
vocal	adj.	大声表达的	U15
volatility	n.	波动	U9
voluntary	adj.	自愿的	U15
vulnerability	n.	漏洞	U9
warehouse	n.	仓库	U10
whereby	adv.	即	U9
wield	v.	行使;运用	U11
workforce	n.	劳动力	U9
yearn	v.	渴望	U10

Terminology List

3PLs (third-party logistics service provider)	n.	第三方物流	U10
activism	n.	行动主义	U12
actuation	n.	驱动	U1
actuator	n.	传动装置	U13
ADC (analog-to-digital converter)	n.	模/数转换器（模数转换器）	U7
added value application	n.	增值应用	U14
addressability	n.	寻址能力	U1
aggregation	n.	聚合	U3
analytics	n.	分析学；解析学	U14
APAC (Asia-Pacific)	n.	亚太地区	U15
apron	n.	停机坪	U8
APUs (auxiliary power units)	n.	辅助动力单元	U8
Arm	n.	安谋国际科技股份有限公司	U10
automation	n.	自动化	U9
backbone network	n.	中枢网络	U3
bandwidth	n.	带宽	U3
barcode scans	n.	条码扫描	U10
beneficiary	n.	受益人	U12
big data	n.	大数据	U12
biometric information	n.	生物信息	U11
Bluetooth	n.	蓝牙技术	U14
broadcast	n.	广播	U3
business intelligence application	n.	商业智慧应用层	U14
call screening	n.	呼叫筛选	U11
CAPSIS (Consolidated Asset Portfolio and Sustainability Information System)	n.	合并资产组合及可持续信息系统	U6
cellular	n.	蜂窝网络	U4
census	n.	人口普查	U12
Cisco	n.	思科公司	U9
client	n.	客户端	U3

code of conduct	n.	行为准则；规范	U15
coil	n.	线圈	U2
communication module	n.	通信模块	U1
configure	v.	配置	U3
connectivity	n.	连通性；互联互通	U10
cost optimization	n.	成本优化	U14
CRM (customer relationship management)	n.	客户关系管理	U10
data custody chain	n.	数据保管链	U15
data format	n.	数据格式	U14
data repository	n.	数据储存库	U14
data-crunching technologies	n.	数据处理技术	U15
data-insight	n.	数据洞察力	U10
DC-DC	n.	DC-DC 转换器（是一种在直流电路中将一个电压值的电能变为另一个电压值的电能的装置）	U7
deploy	v.	部署	U3
deployment path	n.	部署路径	U10
DHS (Department of Homeland Security)	n.	美国国土安全部	U6
directory service	n.	目录服务；查号服务	U14
DLA Piper	n.	欧华律师事务所	U15
dotcom bubble	n.	网络泡沫	U14
drone	n.	无人机	U6
DVR	n.	硬盘录像机	U13
edge-of-network	n.	边缘网络	U9
Edward Snowden	n.	爱德华·斯诺登	U13
EIU (Economist Intelligence Unit)	n.	经济情报所（经济学人集团旗下智囊机构）	U15
EKG (electrocardiogram)	n.	心电图	U7
electromagnetic energy	n.	电磁能	U2
electronic code	n.	电子编码	U11
end user	n.	终端用户	U4
EPC (Electronic Product Code)	n.	电子产品码	U2
ERP system (enterprise resource planning system)	n.	企业资源规划系统	U10
escrow	n.	委托付款服务	U13
Ethernet	n.	以太网	U4
FAA (Federal Aviation Administration)	n.	美国联邦航空管理局	U13
FCC (Federal Communications Commission)	n.	美国联邦通信委员会	U13
FDA (U.S. Food and Drug Administration)	n.	美国食品及药品监督管理局	U13
FEC (Federal Election Commission)	n.	美国联邦选举委员会	U13

feedback loop	n.	反馈回路	U9
frequency	n.	频率	U2
FTC (Federal Trade Commission)	n.	美国联邦贸易委员会	U13
Gartner Magic Quadrant	n.	高德纳魔力象限	U10
gateway	n.	网关	U3
GDPR (General Data Protection Regulation)	n.	通用数据保护条例	U15
geofencing	n.	地理围栏	U8
geotarget	n.	地理定位	U6
GUI (graphical user interface)	n.	图形用户界面	U7
Heathrow Airport	n.	(伦敦)希思罗机场	U8
high-resolution	adj.	高分辨率的	U7
home networking	n.	家庭网络	U5
HomeKit	n.	智能家居	U14
hop	v.	跳跃	U3
i.MX	n.	一种图形处理器	U7
identifier	n.	标识符	U2
IEEE (Institute of Electrical and Electronics Engineers)	n.	电气和电子工程师协会	U5
IETF (Internet Engineering Task Force)	n.	国际互联网工程任务组	U5
Industry 4.0	n.	工业 4.0	U9
industry-wide standard	n.	行业标准	U14
intelligent communication loop	n.	智能通信回路	U9
interface	n.	界面;接口	U14
interlinkage	n.	[计]互连;链接	U15
International Organization for Standardization	n.	国际标准化组织	U15
Internet-connected device	n.	互联网连接设备	U14
interrogator	n.	询问机	U2
Interstate 80 (I-80)	n.	80 号州际公路	U6
IP protocol stack	n.	IP 通信栈	U1
IP solutions	n.	IP 解决方案	U10
iPhone ecosystem	n.	iPhone 生态系统	U14
ITU-T (International Telecommunication Union-Telecommunication Standardization Sector)	n.	国际电信联盟电信标准分局	U5
kilohertz (kHz)	n.	千赫	U2
landside	n.	机场公共场所	U8
lightweight overhead	n.	轻量开销	U9
linear	adj.	线性	U3
load-monitoring	adj.	负荷测录	U10
localization	n.	本地化	U1

location and navigation technologies	n.	定位导航技术	U11
loyalty membership	n.	忠实会员	U10
M2M protocol	n.	M2M 协议	U14
manipulation	n.	操作,操控	U2
market presence	n.	市场形象	U14
Masimo Radical-7	n.	迈心诺 Radical-7 血氧监测仪	U7
McKinsey	n.	麦肯锡	U10
memory-dialing function	n.	记忆拨号功能	U11
mesh	adj.	网状	U3
metadata	n.	元数据	U12
microcontroller	n.	微控制器	U7
microelectronics	n.	微电子学	U1
microwave antenna	n.	微波天线	U2
middleware	n.	中间件	U2
mobility-as-a-service	n.	出行服务	U8
MQTT(message queuing telemetry transport)	n.	消息队列遥测传输	U4
multi-hop routing	v.	多跳路由	U3
multiplicity	n.	多重性	U14
multiprotocol gateway	n.	多协议网关	U14
nanotechnology	n.	纳米技术	U11
NHTSA (National Highway Traffic Safety Administration)	n.	美国高速公路安全管理局	U13
nitrogen dioxide	n.	二氧化氮	U8
node	n.	节点	U3
NSA (National Security Agency)	n.	美国国家安全局	U13
OECD (Organization for Economic Co-operation and Development)	n.	经济合作与发展组织	U15
off-airport transit	n.	机场外交通	U8
olfactory receptor genes	n.	嗅觉受体基因	U11
OMA (Open Mobile Alliance)	n.	开放移动联盟	U4
omnichannel	adj.	多渠道	U10
Op-Amp(operation amplifier)	n.	运算放大器	U7
OPC-UA (object linking and embedding for process control-unified architecture)	n.	用于过程控制的对象链接与嵌入-统一架构	U9
optical fiber	n.	光纤	U3
petabytes	n.	拍字节	U10
PLC (power line communication)	n.	电力线上网	U5
processor	n.	处理机,处理器	U1
product identification	n.	产品识别	U5
proprietary system	n.	专属系统	U12

protocol	n.	协议	U3
proxy	n.	代理	U1
public data-management platform(DMP)	n.	数据管理平台	U10
real-time data	n.	实时数据	U10
redundancy	n.	冗余	U3
RFID (Radio Frequency Identification) infrastructure	n.	无线射频识别基础设施	U1
RFID tags	n.	射频识别标签	U11
rideshare	n.	共乘	U8
ROI (return on investment)	n.	投资回报率	U10
routing	n.	路由	U5
ruggedized	v.	加固抗震的	U10
SCADA (supervisory control and data acquisition)	n.	监控与数据采集	U4
security hole	n.	安全漏洞	U14
sensor	n.	传感器	U11
sensor field	n.	传感域	U3
Shiseido	n.	资生堂（日本公司）	U10
sink node	n.	汇点	U3
smart objects	n.	智能对象	U11
SMEs (subject matter experts)	n.	领域专家	U8
social sorting	n.	社会分类	U11
software framework	n.	软件框架	U4
stack	n.	堆	U14
star	adj.	星形	U3
tagged objects	n.	标记对象	U11
telecommunication	n.	电信	U5
telematics	n.	远程信息处理	U4
telemedicine	n.	远程医学	U14
telemetry	n.	遥测技术；遥感勘测	U4
the right to erasure	n.	删除的权利	U15
the rights to object	n.	反对的权利	U15
time to market	n.	上市时间	U14
time to value	n.	价值实现时间	U10
topology	n.	拓扑	U3
transceiver	n.	收发器	U3
transponder	n.	发射机应答器,询问机,转发器	U2
treasure trove	n.	宝藏,宝库	U10
tree	adj.	树形	U3
ubiquitous networks	n.	泛在网	U5

UI (user interface)	n.	用户界面	U1
UN (United Nation)	n.	联合国	U15
untapped data	n.	未挖掘的数据	U10
usage pattern	n.	使用模式	U14
user terminal	n.	用户终端	U3
utility pole	n.	电线杆	U6
value chain	n.	价值链	U15
video capture	n.	视频采集	U11
voicemail	n.	语音信箱	U11
WAN (wide area network)	n.	广域网	U7
wireless mesh	n.	无线网状网	U4
wireless sensor networks	n.	无线传感器网络	U3
wireless speaker	n.	无线扬声器	U14
WLAN (wireless local area network)	n.	无线局域网	U3
WMAN (wireless metropolitan area network)	n.	无线城域网	U3
WPAN (wireless personal area network)	n.	无线个人网络	U3
WWAN (wireless wide area network)	n.	无线广域网	U3
Zebra Technologies	n.	斑马技术公司	U10
ZigBee	n.	紫蜂	U2